P9-DFU-557

ETHICS IN QUALITY

AUGUST B. MUNDEL

Professional Engineer, Consultant
White Plains, New York

Marcel Dekker, Inc. **New York • Basel • Hong Kong**

ASQC Quality Press **Milwaukee**

Library of Congress Cataloging--in--Publication Data

Mundel, August B.
Ethics in quality/August B. Mundel.
p. cm.
Includes bibliographical references and index.
ISBN 0-8247-8513-4
1. Quality control. 2. Engineering ethics. I. Title.
TS156.M85
620'.0045-- --dc20 91-7176
CIP

Marcel Dekker, Inc.
270 Madison Avenue, New York, New York 10016

American Society for Quality Control
310 West Wisconsin Avenue, Milwaukee, Wisconsin 53203

Current printing (last digit):
10 9 8 7 6 5 4 3 2 1

Printed in the United States of America

Preface

The subject of ethics is one which is not regularly discussed in engineering studies or periodicals. My introduction to ethics, in the context it is treated in this volume, has come about through my working with lawyers on incidents in which a product or service was blamed for a loss or injury to a person or corporation.

An organization's failure to build a safe and reliable product has often resulted in statements to the effect that the company did not behave in an ethical manner. What should have been said is that the organization did not behave in a responsible manner. In fact, many people have stated that the failure to employ adequate quality practices was the cause of the loss sustained by the injured party.

Quality practitioners have thereby become the scapegoats who are blamed for product losses and for personal injuries and losses related to products. It is frequently said that quality control has failed.

Having spent many years in industry, and having observed the quality function and other operations within the corporate structure, I am familiar with the penchant for management, customers, and the public to blame the quality control organization whenever a nonconforming product was found in the field, and when either a conforming or nonconforming product appeared to contribute to any kind of loss. It is with rare exception that you hear the design or manufacturing departments blamed for bad products, or the bad effects of products. Both design and manufacturing, as well as top management, are more frequently the cause of producing and shipping problem-causing product.

In addition, many instances have been observed where management, both middle- and top-level, has taken exception to suggestions for product improvement or delays in shipment to allow for proper corrective action. All of these situations place a heavy ethical responsibility upon the principled engineer, and particularly on ethical quality, reliability, and safety people.

In considering some articles which tend to impugn the ethical behavior of those who supplied a product, I have come to consider what I would have done had I been the engineer, quality manager, or inspector in similar circumstances.

I am proud of some of the decisions that I have made at various times, which I feel protected my fellow employees and the public. I also recall some decisions that I should have made differently.

The proliferation of articles in the public press concerning engineers and quality people suffering dismissal and loss of rank because of their actions, which in some instances was ethical, although their superiors' decisions may have been unethical or dishonest, has led me to an examination of professional societies' codes of ethics and how they are enforced. It is also important that you consider what would be the proper action were you faced with similar difficult situations.

Like many things which the quality engineer examines and encounters, there is seldom a clear-cut situation, or an absolute right and wrong decision. Usually you are faced with decisions that are in the gray area. It is not always clear which decision is the correct one, and two people may honestly believe they are making the correct decision, and yet not agree on what should be done.

The book will try to teach not what is right or wrong, and what is ethical or unethical, but more generally to identify situations where you ought to consider what is ethical and how best to act.

August B. Mundel

Contents

1

Ethics, Quality, and Product Liability

When we choose a product or service, we base our selection on many factors including performance, price, delivery, availability or service, reliability, maintainability, appearance, serviceability, and what we perceive as quality. These factors have led us as a nation to buy many products made overseas. Some years ago we would have bought only American-made products, because at that time we thought that was our best buy. In the years following World War II, we have steadily increased the amount of overseas products we buy.

The Japanese radio and television industry has become so dominant that it is difficult to buy a unit made in the United States. Japanese and European cars have taken over a significant portion of the market. In fields such as consumer electronics, clothing, and some industrial products, it is sometimes difficult to find a domestic product of the type for which we are searching that is economically procurable.

We have come to ask ourselves whether it is ethical to be sending so many jobs overseas. We may also ask if it is possible that the workers of this country are less capable than those overseas. Is the superiority of foreign products related to an ethical problem? Do overseas manufacturers provide well-designed, well-built, and worthwhile products, while domestic manufacturers working with domestic labor find it impossible to supply product of equal quality and worth? Is it due to failure to invest in modern equipment, a financial problem associated with the cost of labor, or is it true that domestic workers are not capable of making an honest product? Do our people and factories lack the ethical factors which contribute to doing a job properly and well? We can, however, do some things very well. The market for the audiophile (high quality audio products) is essentially a U.S. industry.

The world has shrunk. We can see events as they happen in a distant land, in space, and even on the moon. In the past it took weeks for information to travel from one part of the United States to another. The result of this shrinking is that the people in far-off places want many of the same products that we have. This means that they will either make or buy them. In this way the trade increase has been substantial. If they decide to make some of these products, it may well turn out that they recognize that there is a profit to be made by selling some in the United States. If they are less expensive and/or represent more value than the domestic product, they may be very successful. This has the effect of placing pressure on the domestic producers to improve their product. These steps may have a beneficial effect on consumers and at the same time take some jobs out of the domestic economy. In turn there is pressure to convert domestic facilities and labor to other requirements. The economic aspects of this are complex and not the scope of this book. The portion that is within the scope is the ethical problems facing employers.

Employers have a responsibility to try and keep their factories filled so that they can continue to pay themselves, their stockholders, the banks and other sources that have lent them money, their property taxes, and other expenses. To keep the money flowing employers may resort to buying some product components abroad and assembling them here. As a result they may increase their market share and employ more people and increase their business. The variety of results employers may obtain from this import trade include the sale of more product to other countries and the closing of domestic plants. The balance in some industries seems to occur when the foreign product takes over part of the market. In the United States the radio and television market has been taken over by the Asians, the fine camera market largely by the Japanese who have supplanted the Germans, and the motorcycle market also by the Japanese, although one United States producer has established a profitable niche. Some drugs are imported and some are exported. The market shifts and employers have the responsibility of operating profitable operations. It would be safe to say that there is now a world economy rather than a domestic economy.

There is still another question that arises when we examine the marketplace. This is associated with the litigious nature of United States society, and the large number of product liability suits that are filed each year because of deaths, injuries, and losses that result from accidents. There is some opinion that the number of injuries and their severity is associated with products that should not be on the market. There are charges that the workmanship on many products is shoddy and below par, and that domestic manufacturers by placing substandard material on the market, either in design, use, or construction, are not acting in an ethical manner.

This failure of product—whether it be a small trinket, an automobile, an accessory, an appliance, a tool, a vehicle, a building, or a structure such as a bridge or house—is blamed not only on the use of substandard materials, but also on a failure to really care whether the product is properly assembled and tested. The ultimate blame is often placed on the failure of quality control to have performed its job properly and ethically.

In direct opposition to this concept, there is the manufacturer's belief that many of these accidents are caused by careless and senseless product use and abuse. Manufacturers may claim that they provide guards, warnings, and directions which are removed, discarded, and disregarded. This theory says the user was totally responsible for the accident.

The Final Report of the National Commission on Product Safety (June 1970) states:

> Perhaps a case can be made for the acceptability of wilful personal risk-taking by an occasional well-informed consumer, but there is no justification for exposing an entire populace to risks of injury or death which are not necessary and which are not apparent to all. Such hazards must be controlled and limited not at the option of the producer but as a matter of right to the consumer. Many hazards described in this report are unnecessary and can be eliminated without substantially affecting the price to the consumer.
>
> Unfortunately, in the absence of external compulsion it is predictable that there will continue to be an indecent time lag between exposure to a hazard and its elimination. Other advanced nations apparently have discovered this flaw in the output of competitive free enterprise and have made safe products an ongoing governmental objective.
>
> This report proposes means which afford American industry an opportunity to progress voluntarily toward product safety and which, at the same time, should guarantee a new dynamism to that effort. Our suggested procedures are equitable to the consumer and producer alike. The goal is clear. This nation's safety standards and practices must have an exemplary quality consistent with the primacy of American technology [1].

This examination of the ethics of engineering, manufacturing, and quality control has been precipitated by some of the foregoing observations. Surely there are many other areas where we have been compelled to consider the question of ethics. There have been forced mergers, unfriendly takeovers, scandals in government, scandals on Wall Street where people have used inside information to accumulate fortunes, mergers with golden parachutes, and greenmail. When an individual or group accumulates a fortune in one of these mergers, buyouts, or takeovers, it must come from someone else's pocket. In our society it is an acceptable practice to accumulate wealth. But is it ethical to use such devices to accumulate wealth? Do the operations which result in the accumulation of wealth serve the interest of the nation and contribute to the development and improvement of our industrial capabilities? According to a lead article in *Forbes*, November 28, 1988 [2] the incentive for the leveraged buyout is the tax benefit. Interest is tax deductible, but dividends are not. Is such a benefit in the national interest or does it benefit a few?

This is not a rhetorical question. You will be asked again and again throughout this book to make up your mind as to what the proper procedure is, and whether the action was ethical or not. Ethics is seldom a black-and-white situation. An action that might seem ethical in one case may not seem ethical in another, or to another individual.

The code of ethics of one profession is not identical to that of another. Some of these codes will be presented in their entirety; others are much too long, but portions will be considered in this volume. This book is devoted to the ethical behavior of engineers and those employed in kindred professions who come under the general umbrella of engineering, and the industries to which engineers contribute most of their efforts. To this end, several engineering codes are included and discussed. Several other codes will be referred to, but not in as much detail. The differences between ethical, moral, and legal behavior will be discussed. Sometimes the proper ethical and moral behavior for some of us is at odds with what is legal. This provides a dilemma and some of us are likely to find ourselves in difficulty when we try to do what we think is correct. These conditions are becoming more common in politics and business. In politics or government individuals are criticized for being disloyal to their bosses—sometimes a high officer or even the president—when they tell the truth about events. In these instances you must ask, if individuals owe a greater loyalty to their boss, themselves, the public, or their country? The conclusion by different individuals who are rendering judgment in these cases is not always the same. Do codes of ethics make these loyalties clear?

Similar situations occur in industry, government, and contractual relations. This may result in an individual resorting to whistleblowing, going public with stories of cheating, poor quality, overcharging, and shoddy design. We may or may not agree with the individual who engaged in the whistleblowing. We will consider the ethics of whistleblowing and when, if ever, it is the ethical thing to do.

The question of what is ethical and what is unethical comes to the fore when individuals undertake a task. Is it ethical for them to assume that they will be able to do the job well and honestly even though they had not done a similar or related job heretofore? Is it ethical to undertake a job that requires knowledge of a field unfamiliar to you? On the other hand there must always be a bigger bridge, a taller skyscraper, a more novel car design, a change from the old bias cord automobile tire, to the radial tire, and the like. If we did not push past old frontiers there would be no innovation, invention, or improvement.

Does the fact that the job has not been done before, or that individuals have not had specific experience, preclude them from attempting the task or prevent them from performing it with excellence? The question of lack of experience ought to cause workers to ask, is there a high probability that I can successfully perform this task? Only if the answer is yes would it be ethical to undertake the task at someone else's expense.

Return to the problem that was raised a while back, relating to the appeal, appearance, serviceability, and quality of product made here and abroad. You might look into the why and wherefore of the lack of appeal and quality of some products. A quality control professional might correctly ask whether it is really all the fault of poor quality control, and if it is, who is responsible? Is it quality control, the inspector, or someone else? Submitting substandard product for acceptance is certainly not new. In fact, there are examples of the procedures attempted in earlier times, such as during the erection of the Brooklyn Bridge, now over 100 years ago. It is reported that not only was substandard suspension cable wire submitted and rejected, but it was also resubmitted for acceptance without treatment or correction. There are references to similar attempts in antiquity.

Shipping material that does not meet specifications is certainly not a practice that is normally recommended. Still, the practice of shipping material that almost meets specifications is not unusual. The hope is that the material will be suitable for the application and perhaps serve the end use just as well as if it had met specifications. Sometimes there is a procedure for doing just this. The maker and the customer agree that the material can be shipped and used without any effect on the end use. When material is shipped that does not meet the tolerances or specifications and there is no notification on the part of the shipper, is it ethical? Is it ethical to resubmit a shipment to the customer after some work or some sorting has been performed upon the lot? There may be situations where any of these practices is proper, and others in which it is improper. In some situations it may not be legal. The question of what is ethical is often associated with the question of which procedure will cost less or make the most money. Is the adoption of what will be the most economically advantageous practice ethical? The old-fashion engineering and quality concept was that if the product was within the tolerances, it was satisfactory. There is a newer concept that product should be close to target and that the closer to target it is the greater the value. The material out near the tolerances is of less value. As a result, there is pressure on manufacturers and quality personnel to add as much value to the product as possible by better statistical process control, and to deliver the best product possible. This often results in more user acceptance and less cost. One responsibility of the quality personnel is to add value and reduce cost. Ethics requires the exercise of these responsibilities.

In the context of the last paragraph, the actions of manufacturers or shippers may be expeditious, but there may be a question as to whether they are legal. In this context *legal* means that there is no law that specifically inhibits the action. The term *ethical* conveys the concept of it being fair and in accordance with the intent of the agreement.

Sometimes when we think of ethics we come away with the idea that the correct practice is going to cost more than the unethical practice. Ethics thus becomes confused with economics. Examples of situations where this occurs will be discussed, and again you will have to make up your mind as to whether one practice is ethical and another not. An excellent way to really get involved is to look for everyday life situations and decide how an individual thinks he or she would react if faced with the same question that the other person faced. Would your decision have been the same as his or hers?

Is it ethical to pocket money one finds on the street? One cent? Five cents? Ten cents? One dollar? One hundred dollars? One thousand dollars? Ten thousand dollars? A million dollars? How big does the sum have to be before you start looking for an owner?

For those in quality control there are decisions to be made every day. Are all the decisions truly ethical, or are there slight adjustments made because the boss would like things done his or her way? Perhaps a boss even insists on ways specifications should be interpreted.

Before we can really decide what is ethical or not, we must adopt some definitions so that ethics can be examined from the same point of view. These are strictly dictionary definitions. They describe what is generally meant by ethics and morals and illegal or criminal acts, but they do not describe what the real differences are between what is ethical for one person and not for another. Because members of different professions place different emphasis on practices, what is ethical to one might not be to another.

For the purposes of the presentations, discussions, and questions that are posed the following terms are defined: according to the *American Heritage Dictionary of the English Language*, Ethics and Morality are:

Ethics: The rules or standards governing the conduct of the members of a profession.

Morality: A set of customs of a given society, class, or social group, which regulate relationships and prescribe modes of behavior to enhance the group's survival [3].

The following general definitions will also be useful:

Product liability: The legally imposed responsibility for injury or loss suffered as a result of use, contact with, or performance of a product or service. Any time that there is an injury or a loss associated with contact or the result of some incident relating to a product and/or service and there is a claim for restitution, compensation or recovery the incident will be classed as relating to product liability.

Tort: A wrongful act, resulting in a loss or injury compensable by a product liability verdict or settlement.

Criminal act: An action in violation of the criminal code as distinguished from the civil code. It can include a willful act, done with malicious intent to injure or harm an individual or his property.

Illegal act: An action prohibited by law.

The question of whether something is illegal may or may not be cause for considering the act ethical or unethical. As an example one might consider the fact that slavery was legal at one time. How one could treat or mistreat slaves may also have been written into various laws. This did not affect the ethics of slavery.

2

Engineering Ethics

The ethical decision is one which conforms to the rules of conduct of members of a profession. Rules would never be made if there were not some instances where individuals acted in a manner that others thought was wrong. It is not always easy to make ethical decisions. This is particularly so when the ethical course costs more.

The one lesson of quality control which usually surprises both manufacturing and financial people is that, usually, the better the quality of the product, the less it costs to produce. Although this may seem amazing, it is true. Quality means conformance. Some of the costs of nonconformance are due to the decreased yield of good salable product, the costs of rework, repair, replacement in the field, and lost sales.

If manufacturers start out delivering products that do not conform, several questions arise:

- Should the products be recalled and exchanged or corrected, or should refunds be made?
- Should manufacturers just forget the whole mess and let customers complain?
- If customers complaints, should manufacturers correct the products at their expense, or charge for the correction?

What is the ethical practice to adopt? What is the cost of the ethical decision? If the problem could have been avoided by correction and/or redesign before manufacture, then the costs might have been less. It is also possible that, were the correct design used, the product would cost more to produce. Under these conditions the decision to use a more expensive material might be less attractive, even if it was clearly the proper and the ethical decision. On the other hand, a satisfactory design might have resulted in a lower total cost. It must always be remembered that there are many ways to design a product, or do a job. A more expensive one may be better than what is at hand. This does not prove that there is not another way to do the job that costs less and is even better. Sometimes brainstorming sessions, or similar techniques, can resolve these problems.

Many believe ethical and product liability problems faced by companies with large liability losses are due to the consumer movement dating to the late 1960s. It is true that in earlier times it was more difficult for people in the United States to recover for injuries or losses. Some may say that the consumer revolution caused us to become more litigious. Others may say that lawyers had not learned how to recover and profit from such litigation. Some may even say that fewer people were injured. There are no good answers.

The concept of trying to recover when one is injured is not new. In 1758 B.C., Hammurabi, King of Babylon, proclaimed a series of laws which can be found on an iron column now in The Louvre in Paris. One of these laws concerned builders whom he hoped would help in the expansion of his realm.

> If a builder has built a house for a man and has not made his work sound, and the house which he has built has fallen down and so caused the death of the householder, that builder shall be put to death. If it causes the death of the householder's son, they shall put that builder's son to death. If it causes the death of the householder's slave, he shall give slave for slave to the householder. If it destroys property he shall replace anything it has destroyed; and because he has not made sound the house which he has built and it has fallen down, he shall rebuild the house which has fallen down from his own property. If a builder has built a house for a man and does not make his work perfect and the wall bulges, that builder shall put the wall into sound condition at his own cost [4].

The recompense for an ill-built house certainly was not forthcoming if a man lost his life or his son. If he lost a slave, he was repaid, and the house had to be put right. The law of Hammurabi was a forerunner of some modern product liability laws. Unlike most laws it established specific payments and penalties. Most ethical codes of conduct for professionals do not provide for specific punishment.

Ethical codes are promulgated by professional groups which have less power of enforcement than Hammurabi. An early ethical code is the Hippocratic oath. It is reported that this was adopted by a medical group, and that other professional groups attempting to copy the medical profession adopted ethical codes. The Hippocratic oath is administered to many medical school graduates in the form shown (Figure 2-1). It is a much amended version of the principles of Hippocrates, who lived about 400 B.C..

The terms of the oath are not onerous. It is an agreement to work for the aid of the patient rather than primarily for the doctor's benefit. It also requests that the secrets of the patient will be held inviolate. The one who administers the oath wishes prosperity and good repute provided doctors practice their art in compliance with the oath.

Fundamentally the oath leaves out some controversial statements. Do doctors owe a greater loyalty to their patients or their country? There are situations where physicians are required to report injuries and diseases to authorities. Perhaps when the situation is analyzed, there is a difference between loyalty to country and to government.

In the engineering profession there are other suggestions and agreements we make when we sign an application to join one of the engineering societies or become a member of a profession. There is the acceptance of responsibility for public safety.

One of the specific prohibitions of the Hippocratic oath is the agreement to "perform no operation for a criminal purpose, even if solicited, far less to suggest it." It might be inferred from this prohibition that it was a far greater violation of the code to initiate the prohibited operation than to perform it upon the suggestion of others. Perhaps that explains the excuse some have proffered, "The devil made me do it!"

Legally there may be no distinction as to who initiated the idea to commit murder. All those involved seem equally guilty insofar as morality or ethics is concerned. Murderers must recognize what they do and how they become involved, rather than claim that they are innocent because the crime was someone else's idea.

The ethical codes of different professions are quite varied. In fact, engineering and technical society codes are not identical. Emphasis is placed on different aspects of behavior. There are also codes of ethics for government service.

The code of ethics for attorneys is covered either by the Model Code of Professional Responsibility and Code of Judicial Conduct, or the Model Rules of Professional Conduct and Code of Judicial Conduct. These are booklets of approximately 130 and 150 pages respectively. The reason for two volumes is that all states have not adopted the later version.

Within states there are annotated versions with judicial constructions from the courts and state bar associations. In the state of New York McKinney's Book 29 of Judiciary Law contains 740 pages, plus addenda.

Attorneys are officers of the court who also have special duties to their clients. This is not to say that they are not responsible to the public. However, clients have the right to the presumption of innocence in a criminal matter and the right to an attorney. Attorneys' tasks are to act as legal representatives with their first duty to try and win the case for their clients.

There are conflicts. Sometimes an attorney is able to obtain a verdict of innocence for a client, even though he or she may know that the client committed the act. This is not unethical, but it does not seem to be in the public interest.

Attorneys may withdraw from a case under specific conditions, which would include unethical behavior on the part of a client. The state codes spell out many of these examples. In Florida a man, under the influence of a substance, was struck and killed by a motorist who left the scene of the accident without identifying himself. He reported the incident to a lawyer, who in turn was not required to reveal the motorist's identity. To some of us, this may seem unethical, but the law ruled otherwise.

Physicians have as their first duty the health and protection of the life of their clients. They also have a duty to report instances of specific diseases and gunshot wounds.

Professionals cannot always practice in any state, since the laws differ. Attorneys are usually limited to practice in federal courts or in the courts of the state where they are licensed. Physicians, dentists, and engineers are also licensed by the states. In some instances a practice cannot extend beyond state lines. Engineers must usually obtain a license in any state where they maintain an office.

Engineers, attorneys, and other professional may maintain licensure in several states. The activity of working for a manufacturing company as an engineer is not considered to be the practice of professional engineering. This allows engineers to pursue a career without being licensed.

The rules for obtaining and maintaining an engineering license are quite specific. To some degree they differ state by state, but in general younger engineers are required to have graduated from accredited institutions and to pass examinations. Some, with more experience and expertise, can become licensed by application. The maintenance of the license requires an annual or biennial fee. Once licensed in a state, a license in another state is usually obtainable by application, rather than examination.

The state of California licenses Professional Engineers by category—electrical, mechanical, civil, structural, chemical, quality, and so on. When a new category is created, it usually has a grandfathering clause. All who meet the qualifications can be licensed by application. This is easier than the emeritus requirement or the examination procedure.

Some individuals have obtained a first license in a specialty in California under a grandfathering clause, and then used the reciprocity agreement to obtain a license in another state. Most states license engineers as Professional Engineers. There is no category or restriction. Engineers are responsible for restricting themselves to fields where they are competent. The parlaying of a specialty license (as issued in California) into a general license is apparently quite proper. To practice in a field where one is not qualified is not proper.

If engineers, who have parlayed their California licenses in quality engineering into general licenses in engineering, sign structural plans are they acting ethically?

I obtained my engineering license in New York state some years back. I took the examinations in structures, hydraulics, statics, electrical engineering, and economics. A lawyer admitted to the California bar and familiar with the California law asked me what limitations were placed on my license. I told him none. He then asked if I could sign structural plans, which must be filed for buildings, and I admitted that I could. He then asked what gave me that right. Actually, it was the license, but I told him I had passed the same structural examination that civil or structural engineers took. This is apparently not true in all states. Ethically I believe I can sign any set of plans, or undertake any work I feel competent to perform.

The lawyer then asked if, in fact, I did sign structural documents. I said that, if they were simple, I did. A complicated structure would take me much longer than it would take one who designs structures on an everyday basis. For economy I would recommend that the client use the services of an associate who works on structures regularly.

A doctor told me he had to relearn medicine several times. First when the sulfa drugs were introduced, and then when penicillin and other antibiotics were developed. Engineers also must learn modern techniques and how to use modern materials and equipment.

Most good engineers and other professionals abide by the seventh Fundamental Canon of the Accreditation Board for Engineers and Technology (ABET) that requires continuation of professional development throughout their careers. There are unfortunately many exceptions.

It is difficult to estimate the fraction of engineers or other quality professionals who have really continued their professional development. Many engineering graduates leave the profession for other activities. The largest engineering society, the Institute of Electrical and Electronics Engineers, Inc. (IEEE) has approximately 300,000 members. The next several societies have fewer. The total of the top 20 probably is in the order of 800,000, yet U.S. engineering schools graduate 50,000 bachelor level students per year. The average 16-year life of society members indicates a gradual dropout of individuals from the profession.

It is true that not all engineering graduates join engineering, scientific or technical societies. Some go on to careers in medicine, law, business, and other activities. Some work as engineers, but do not join a society. On the other side of the ledger are those who become members of professional societies without having graduated from a school of engineering. The American Society of Quality Control (ASQC) is an example of such a society where graduation with a recognized degree from college or higher education is not required.

One advantage of joining an engineering society, or ASQC, is that it provides periodicals with articles on modern techniques, and helps engineers learn more of modern theory and technique. It also helps to maintain a forum for such information and thereby aids others and contributes to progress.

The further education of some engineers is so neglected that in the course of a few years their skills are less than that of current graduates. These young veterans then become much less useful.

Here the failure to make an effort to conform to ethical behavior results in the termination of usefulness and the ability to earn a reasonable livelihood. In the engineering profession this is one of the most severe penalties that society imposes.

There are other penalties that can be imposed for lack of conformance to ethical codes. The medical and legal professions appear to be more active in enforcing their ethical codes. There may be some argument as to how energetic these groups are. Nonetheless, I believe that there is more activity in trying to uphold ethical standards in law and medicine than in engineering, although activity in some areas seems to be extremely slow. The apparent difference may be due to the greater publicity given these professionals, or their exposure to pressures for unethical practice, since many practice as individuals rather than as employees of organizations.

The engineering societies I have contacted, with exceptions, have had few instances where members have been ousted or suspended for violations of the society's code of ethics. In many instances losing one's society membership due to ethical failures need not cause the loss of professional license, if one has been issued. In law and medicine the societies would probably make every effort to have the individual barred from practicing in the state.

One of the most serious cases I recall concerned the engineer/architect responsible for the design of the skywalk across the Kansas City Grand Hyatt ballroom; his license was suspended.

Some years back several members of ASQC led a drive to have a member brought up on charges of unethical use of the society's name due to his use of Quality Control, in his business name. There was not only insufficient support for this claim, but also many more members who believed the individual was not doing anything improper. If there had been a hearing, and the ethics committee had decided in favor of terminating or suspending his membership, this would probably have had little effect on his business.

There is unfortunately little that many of the societies can do to remove from their rolls members who are unethical in one way or another. Infractions are often not reported to the society's committee on ethics, and when they are reported, some committees are loath to act on the allegations. In some instances those who stand accused, or have been suspended by the committee, have sued committee members, causing them to spend large sums to defend themselves.

There is, however, a more serious problem that arises from ethical behavior as some practitioners interpret it. There have been numerous instances, many covered in the public press, where individuals have become whistleblowers. This comes about from a conflicting set of ethical behaviors set forth in most codes.

The Code of Ethics for members of the American Society for Quality Control enjoins a member to be ". . . honest and impartial, and serve with devotion his employer, his clients, and the public" [6].

This phrase, along with a sense of personal integrity, has led some members of our profession into a position of conflict. Not only have they found product, workmanship, or practices not in conformance with specifications, but also, in their opinion, unsafe. Thus they have indicated the work should be rejected or corrective action initiated. When asked to reconsider, they have refused. When the supervisor overruled them, they have gone over the supervisor's head, and in some cases to public authorities, or to the public.

This has resulted in the individual being classed as a troublemaker or whistleblower. Those resorting to these practices have been harassed, terminated, or had their jobs downgraded. Some may say whistleblowers represent a small minority. The frequency of these events is sufficient in government activities to result in congressional action. After all, is it not general opinion that unethical behavior is much less frequent than ethical behavior?

Some of the harassed whistleblowers have resorted to court action, and have been vindicated. Despite this they may still be harassed. Several cases will be discussed in Chapter 9.

Is the whistleblower behaving ethically? Is there a duty and responsibility that overrides that to the employer if it is believed that public safety is not being protected? Is there a similar duty if it is believed that honest costing is not being rendered and superiors will not attempt to correct the situation?

The codes of some societies are specific in encouraging integrity and responsibility to the public for its safety. These codes cause the same inner conflicts for employees who are convinced that their employer, in overruling their professional judgment, is very likely to have an adverse effect on the safety and well-being of the public (ASQC Code, Section 2.3).

ABET is more specific in stating in its first Fundamental Canon: "Engineers shall hold paramount the safety, health and welfare of the public in the performance of their professional duties" [7]. The American Society of Civil Engineers (ASCE) notes in their Code of Ethics that they have adapted the tenets of ABET.

The IEEE is also specific in item 1 of Article IV of their Code of Ethics. "Members shall, in fulfilling their responsibilities to the community: Protect the safety, health and welfare of the public and speak out against abuses in these areas affecting the public interest" [8].

IEEE went to court in the role of amicus curiae or friend of the court in the case of three Bay Area Rapid Transit engineers who were fired and blackballed in a whistleblowing case. See Chapter 9 for details.

Currently there are several cases in the courts relating to whistleblowing and the mental anguish and punishment meted out. One concerns a Morton Thiokol engineer, who claims he tried to stop the January 1986 *Challenger* launch, because there was an indication that the O-ring seals would not perform satisfactorily at the low temperature that existed at Cape Canaveral prior to the launch. The *Washington Post* reported on September 3, 1988 that U.S. District Judge David K. Winder dismissed with prejudice, many counts in the suits filed by Roger Boisjoly against Morton Thiokol Inc. However, two counts of one suit which deals with defamation and conspiracy were dismissed without prejudice. These can be refiled.

The ethical principles that one is urged to support by the codes pose serious problems to the few who find themselves in the unenviable position of being unable to convince their supervisors and management that they have correctly assessed the situation, and that there is great danger in shipping the product or failing to take major corrective actions. What would you have done in similar circumstances?

Fortunately most of us have never had to make a really tough decision on speaking out and taking a public stand opposite that of our employers or clients. Nonetheless, it is possible that some of us have come really close to a similar situation. We have either been able to resolve the question, or rationalize that the situation was not sufficiently critical to force us to take a stand.

On the other hand, would you agree that there is a large segment of the engineering community that has consistently failed to adhere to the less controversial ethical principles which follow?

The ASQC code, in its Fundamental Principles, provides that each member:

- I Will be honest and impartial, and serve with devotion his employer, his clients, and the public.
- II Will strive to increase the competence and prestige of his profession.
- III Will use his knowledge and skill for the advancement of human welfare, and in promoting the safety and reliability of products for public use.
- IV Will earnestly endeavor to aid the work of the Society [9].

The Fundamental Canons of ABET state that:

2. Engineers shall perform services only in the areas of their competence.
7. Engineers shall continue their professional development throughout their careers and shall provide opportunities for the professional development of those engineers under their supervision [10].

The IEEE Code, Article IV, requires that members shall, in fulfilling their responsibilities to the community:

1. Protect the safety, health and welfare of the public and speak out against abuses in these areas affecting the public interest;
2. Contribute professional advice, as appropriate, to civic, charitable, or other nonprofit organizations;
3. Seek to extend public knowledge and appreciation of the profession and its achievements [11].

Some of these responsibilities appear to be accepted as duties by a fraction of the engineers who belong to our societies. Do we really exercise all the effort we should? Are all but a few of us ethical in all respects? Are we, as individuals, really living up to the Code of Ethics to which we subscribed when we joined our technical societies? Are we continuing our professional development? Are we contributing our professional advice to civic and nonprofit organizations? Are we contributing to the well-being of our professional organizations? Do we speak out against abuses?

The Code of Ethics for Government Service is also included in this section so that comparisons can be made between the Government Code and that of engineers and other professionals.

Figure 2.1: HIPPOCRATIC OATH

You do solemnly swear, each man by whatever he holds most sacred, that you will be loyal to the profession of medicine and just and generous to its members; that you will lead your lives and practice your art in uprightness and honor; that into whatsoever house you shall enter, it shall be for the good of the sick to the utmost of your power, you holding yourself far aloof from wrong, from corruption, from the tempting of others to vice; that you will exercise your art solely for the cure of your patients and will give no drug, perform no operation for a criminal purpose, even if solicited, far less suggest it; that whatsoever you shall see or hear of the lives of men which is not fitting to be spoken, you will keep inviolably secret. These things do you swear. Let each man bow the head in sign of acquiescence. And now, if you will be true to this, your oath, may prosperity and good repute be ever yours; the opposite, if you shall prove yourself foresworn.

(From *The New Columbia Encyclopedia*, W. H. Harris and J. S. Levy, eds. [New York: Columbia University Press, 1975], p.1256.)

Figure 2.2: CODE OF ETHICS FOR MEMBERS OF THE AMERICAN SOCIETY FOR QUALITY CONTROL

Fundamental Principles

Each member of the Society, to uphold and advance the honor and dignity of the profession, and in keeping with high standards of ethical conduct:

I. Will be honest and impartial, and serve with devotion his employer, his clients and the public.
II. Will strive to increase the competence and prestige of the profession.
III. Will use his knowledge and skill for the advancement of human welfare, and in promoting the safety and reliability of products for public use.
IV. Will earnestly endeavor to aid the work of the Society.

Relations With the Public

1.1 Each Society member will do whatever he can to promote the reliability and safety of all products that come within his jurisdiction.
1.2 He will endeavor to extend public knowledge of the work of the Society and its members that relates to the public welfare.
1.3 He will be dignified and modest in explaining his work and merit.
1.4 He will preface any public statements that he may issue by clearly indicating on whose behalf they are made.

Relations With Employers and Clients

2.1 Each Society member will act in professional matters as a faithful agent or trustee for each employer or client.
2.2 He will inform each client or employer of any business connections, interests or affiliations which might influence his judgment or impair the equitable character of his services.
2.3 He will indicate to his employer or client the adverse consequences to be expected if his professional judgment is overruled.
2.4 He will not disclose information concerning the business affairs or technical processes of any present or former employer or client without his consent.

2.5 He will not accept compensation from more than one party for the same service without the consent of all parties. If employed, he will engage in supplementary employment of consulting practice only with the consent of his employer.

Relations With Peers

3.1 Each member of the Society will take care that credit for the work of others is given to those to whom it is due.
3.2 He will endeavor to aid the professional development and advancement of those in this employ or under his supervision.
3.3 He will not compete unfairly with others. He will extend his friendship and confidence to all associates and those with whom he has business relations.

(Reprinted with permission of the ASQC.)

Figure 2.3: CODE OF ETHICS AMERICAN SOCIETY OF CIVIL ENGINEERS

Four Fundamental Principles*

Engineers uphold and advance the integrity, honor and dignity of the engineering profession by:

I. using their knowledge and skill for the enhancement of human welfare;
II. being honest and impartial and serving with fidelity the public, their employers and clients;
III. striving to increase the competence and prestige of the engineering profession; and
IV. supporting the professional and technical societies of their disciplines.

Seven Fundamental Canons

1. Engineers shall hold paramount the safety, health and welfare of the public in the performance of their professional duties

ASCE Guidelines:

a. Engineers shall recognize that the lives, safety, health and welfare of the general public are dependent upon engineering judgments, decisions and practices incorporated into structures, machines, products, processes and devices.
b. Engineers shall approve or seal only those design documents, reviewed or prepared by them, which are determined to be safe for public health and welfare in conformity with accepted engineering standards.
c. Engineers whose professional judgment is overruled under circumstances where the safety, health and welfare of the public are endangered shall inform their clients or employers of the possible consequences.
d. Engineers who have knowledge or reason to believe that another person or firm may be in violation of any of the provisions of Canon I shall present such information to the proper authority in writing and shall cooperate with the proper authority in

* The American Society of Civil Engineers adopted THE FUNDAMENTAL PRINCIPLES of the Code of Ethics of Engineers as accepted by the Engineers Council for Professional Development (Renamed Accreditation Board for Engineering and Technology).

furnishing such further information or assistance as may be required.

e. Engineers should seek opportunities to be of constructive service in civic affairs and work for the advancement of the safety, health and well-being of their communities.
f. Engineers should be committed to improving the environment to enhance the quality of life.

2. Engineers shall perform services only in areas of their competence

ASCE Guidelines:

a. Engineers shall undertake to perform engineering assignments only when qualified by education or experience in the technical field of engineering involved.
b. Engineers may accept an assignment requiring education or experience outside of their own fields of competence, provided their services are restricted to those phases of the project in which they are qualified. All other phases of such project shall be performed by qualified associates, consultants, or employees.
c. Engineers shall not affix their signatures or seals to any engineering plan or document dealing with subject matter in which they lack competence by virtue of education or experience, or to any such plan or document not reviewed or prepared under their supervisory control.

3. Engineers shall issue public statements only in an objective and truthful manner

ASCE Guidelines:

a. Engineers should endeavor to extend the public knowledge of engineering, and shall not participate in the dissemination of untrue, unfair or exaggerated statements regarding engineering.
b. Engineers shall be objective and truthful in professional reports, statements, or testimony. They shall include all relevant and pertinent information in such reports, statements, or testimony.
c. Engineers, when serving as expert witnesses, shall express an engineering opinion only when it is founded upon adequate knowledge of the facts, upon a background of technical competence, and upon honest conviction.

d. Engineers shall issue no statements, criticisms, or arguments on engineering matters which are inspired or paid for by interested parties, unless they indicate on whose behalf the statements are made.
e. Engineers shall be dignified and modest in explaining their work and merit, and will avoid any act tending to promote their own interests at the expense of the integrity, honor and dignity of the profession.

4. Engineers shall act in professional matters for each employer or client as faithful agents or trustees, and shall avoid conflicts of interest

ASCE Guidelines:

a. Engineers shall avoid all known or potential conflicts of interest with their employers or clients and shall promptly inform their employers or clients of any business association, interests, or circumstances which could influence their judgment or the quality of their services.
b. Engineers shall not accept compensation from more than one party for services on the same project, or for services pertaining to the same project, unless the circumstances are fully disclosed to, and agreed to, by all interested parties.
c. Engineers shall not solicit or accept gratuities, directly or indirectly, from contractors, their agents, or other parties dealing with their clients or employers in connection with work for which they are responsible.
d. Engineers in public service as members, advisors, or employees of a governmental body or department shall not participate in considerations or actions with respect to services solicited or provided by them or their organization in private or public engineering practice.
e. Engineers shall advise their employers or clients when, as a result of their studies, they believe a project will not be successful.
f. Engineers shall not use confidential information coming to them in the course of their assignments as a means of making personal profit if such action is adverse to the interests of their clients, employers or the public.
g. Engineers shall not accept professional employment outside of their regular work or interest without the knowledge of their employers.

5. Engineers shall build their professional reputation on the merit of their service and shall not compete unfairly with others

ASCE Guidelines:

a. Engineers shall not give, solicit or receive, either directly or indirectly, any commission, political contribution, or a gift or other consideration in order to secure work, exclusive of securing salaried positions through employment agencies.
b. Engineers should negotiate contracts for professional services fairly and on the basis of demonstrated competence and qualifications for the type of professional service required.
c. Engineers shall not request, propose or accept professional commissions on a contingent basis under circumstances in which their professional judgments may be compromised.
d. Engineers shall not falsify or permit misrepresentation of their academic or professional qualifications or experience.
e. Engineers shall give proper credit for engineering work to those to whom credit is due, and shall recognize the proprietary interests of others. Whenever possible, they shall name the person or persons who may be responsible for designs, inventions, writings or other accomplishments.
f. Engineers may advertise professional services in a way that does not contain self-laudatory or misleading language or is in any other manner derogatory to the dignity of the profession. Examples of permissible advertising are as follows:

   Professional cards in recognized, dignified publications, and listings in rosters or directories published by responsible organizations, provided that the cards or listings are consistent in size and content and are in a section of the publication regularly devoted to such professional cards.

   Brochures which factually describe experience, facilities, personnel and capacity to render service, providing they are not misleading with respect to the engineer's participation in projects described.

   Display advertising in recognized dignified business and professional publications, providing it is factual, contains no laudatory expressions or implications and is not misleading with respect to the engineer's extent of participation in projects described.

A statement of the engineers' names or the name of the firm and statement of the type of service posted on projects for which they render services.

Preparation or authorization of descriptive articles for the lay or technical press, which are factual, dignified and free from laudatory implications. Such articles shall not imply anything more than direct participation in the project described.

Permission by engineers for their names to be used in commercial advertisements, such as may be published by contractors, material suppliers, etc., only by means of a modest, dignified notation acknowledging the engineers' participation in the project described. Such permission shall not include public endorsement of proprietary products.

g. Engineers shall not maliciously or falsely, directly or indirectly, injure the professional reputation, prospects, practice or employment of another engineer, or indiscriminately criticize another's work.

h. Engineers shall not use equipment, supplies, laboratory or office facilities of their employers to carry on outside private practice without the consent of their employers.

6. Engineers shall act in such a manner as to uphold and enhance the honor, integrity, and dignity of the engineering profession

ASCE Guidelines:

a. Engineers shall not knowingly act in a manner which will be derogatory to the honor, integrity or dignity of the engineering profession or knowingly engage in business or professional practices of a fraudulent, dishonest or unethical nature.

7. Engineers shall continue their professional development throughout their careers, and shall provide opportunities for the professional development of those engineers under their supervision

ASCE Guidelines:

a. Engineers should keep current in their specialty fields by engaging in professional practice, participating in continuing education courses, reading in the technical literature, and attending professional meetings and seminars.

b. Engineers should encourage their engineering employees to become registered at the earliest possible date.
c. Engineers should encourage engineering employees to attend and present papers at professional and technical society meetings.
d. Engineers shall uphold the principle of mutually satisfying relationships between employers and employees with respect to terms of employment including professional grade descriptions, salary ranges, and fringe benefits.

Figure 2.4: AMERICAN MEDICAL ASSOCIATION PRINCIPLES OF MEDICAL ETHICS (1980)

Preamble:

The medical profession has long subscribed to a body of ethical statements developed primarily for the benefit of the patient. As a member of this profession, a physician must recognize responsibility not only to patients, but also to society, to other health professionals, and to self. The following Principles adopted by the American Medical Association are not laws, but standards of conduct which define the essentials of honorable behavior for the physician.

I. A physician shall be dedicated to providing competent medical service with compassion and respect for human dignity.

II. A physician shall deal honestly with patients and colleagues, and strive to expose those physicians deficient in character or competence, or who engage in fraud or deception.

III. A physician shall respect the law and also recognize a responsibility to seek changes in those requirements which are contrary to the best interests of the patient.

IV. A physician shall respect the rights of patients, of colleagues, and of other health professionals, and shall safeguard patient confidences within the constraints of the law.

V. A physician shall continue to study, apply and advance scientific knowledge, make relevant information available to patients, colleagues, and the public, obtain consultation, and use the talents of other health professionals when indicated.

VI. A physician shall, in the provision of appropriate patient care, except in emergencies, be free to choose whom to serve, with whom to associate, and the environment in which to provide medical services.

VII. A physician shall recognize a responsibility to participate in activities contributing to an improved community.

(Reprinted with permission of the American Medical Association.)

Figure 2.5: ABET CODE OF ETHICS OF ENGINEERS AND SUGGESTED GUIDELINES FOR USE WITH THE FUNDAMENTAL CANONS OF ETHICS

The Fundamental Principles

Engineers uphold and advance the integrity, honor and dignity of the engineering profession by:

I. using their knowledge and skill for the enhancement of human welfare;

II. being honest and impartial, and serving with fidelity the public, their employers and clients;

III. striving to increase the competence and prestige of the engineering profession; and

IV. supporting the professional and technical societies of their disciplines.

The Fundamental Canons

1. Engineers shall hold paramount the safety, health and welfare of the public in the performance of their professional duties.
2. Engineers shall perform services only in the areas of their competence.
3. Engineers shall issue public statements only in an objective and truthful manner.
4. Engineers shall act in professional matters for each employer or client as faithful agents or trustees, and shall avoid conflicts of interest.
5. Engineers shall build their professional reputation on the merit of their services and shall not compete unfairly with others.
6. Engineers shall act in such a manner as to uphold and enhance the honor, integrity and dignity of the profession.
7. Engineers shall continue their professional development throughout their careers and shall provide opportunities for the professional development of those engineers under their supervision.

Suggested Guidelines For Use With The Fundamental Canons of Ethics

1. Engineers shall hold paramount the safety, health and welfare of the public in the performance of their professional duties.

a. Engineers shall recognize that the lives, safety, health and welfare of the general public are dependent upon engineering judgments, decisions and practices incorporated into structures, machines, products, processes and devices.

b. Engineers shall not approve nor seal plans and/or specifications that are not of a design safe to the public health and welfare and in conformity with accepted engineering standards.

c. Should the Engineers' professional judgment be overruled under circumstances where the safety, health, and welfare of the public are endangered, the Engineers shall inform their clients or employers of the possible consequences and notify other proper authority of the situation, as may be appropriate.

   (c.1) Engineers shall do whatever possible to provide published standards, test codes and quality control procedures that will enable the public to understand the degree of safety or life expectancy associated with the use of the design, products and systems for which they are responsible.

   (c.2) Engineers will conduct reviews of the safety and reliability of the design, products or systems for which they are responsible before giving their approval to the plans for the design.

   (c.3) Should Engineers observe conditions which they believe will endanger public safety or health, they shall inform the proper authority of the situation.

d. Should Engineers have knowledge or reason to believe that another person or firm may be in violation of any of the provisions of these Guidelines, they shall present such information to the proper authority in writing and shall cooperate with the proper authority in furnishing such further information or assistance as may be required.

e. Engineers should seek opportunities to be of constructive service in civic affairs and work for the advancement of the safety, health and well-being of their communities.

   (d.1) They shall advise proper authority if an adequate review of the safety and reliability of the products or systems has not been made or when the design imposes hazards to the public through its use.

   (d.2) They shall withhold approval of products or systems

when changes or modifications are made which would affect adversely its performance insofar as safety and reliability are concerned.

f. Engineers should be committed to improving the environment to enhance the quality of life.

2. Engineers shall perform services only in areas of their competence.

a. Engineers shall undertake to perform engineering assignments only when qualified by education or experience in the specific technical field of engineering involved.

b. Engineers may accept an assignment requiring education or experience outside of their own fields of competence, but only to the extent that their services are restricted to those phases of the project in which they are qualified. All other phases of such project shall be performed by qualified associates, consultants, or employees.

c. Engineers shall not affix their signatures and/or seals to any engineering plan or document dealing with subject matter in which they lack competence by virtue of education or experience, nor to any such plan or document not prepared under their direct supervisory control.

3. Engineers shall issue public statements only in an objective and truthful manner.

a. Engineers shall endeavor to extend public knowledge, and to prevent misunderstandings of the achievements of engineering.

b. Engineers shall be completely objective and truthful in all professional reports, statements, or testimony. They shall include all relevant and pertinent information in such reports, statements, or testimony.

c. Engineers, when serving as expert or technical witnesses before any court, commission, or other tribunal, shall express an engineering opinion only when it is founded upon adequate knowledge of the facts in issue, upon a background of technical competence in the subject matter, and upon honest conviction of the accuracy and propriety of their testimony.

d. Engineers shall issue no statements, criticisms, nor arguments on engineering matters which are inspired or paid for by an interested party, or parties, unless they have prefaced their comments by explicitly identifying themselves, by disclosing the identities of the party or parties on whose behalf they are speaking, and by revealing the existence of any pecuniary interest they may have in the instant matters.

e. Engineers shall be dignified and modest in explaining their work and merit, and will avoid any act tending to promote their own interests at the expense of the integrity, honor and dignity of the profession.

4. Engineers shall act in professional matters for each employer or client as faithful agents or trustees, and shall avoid conflicts of interest.

a. Engineers shall avoid all known conflicts of interest with their employers or clients and shall promptly inform their employers or clients of any business association, interests, or circumstances which could influence their judgment or the quality of their services.

b. Engineers shall not knowingly undertake any assignments which would knowingly create a potential conflict of interest between themselves and their clients or their employers.

c. Engineers shall not accept compensation, financial or otherwise, from more than one party for services on the same project, nor for services pertaining to the same project, unless the circumstances are fully disclosed to, and agreed to, by all interested parties.

d. Engineers shall not solicit nor accept financial or other valuable considerations, including free engineering designs, from material or equipment suppliers for specifying their products.

e. Engineers shall not solicit nor accept gratuities, directly or indirectly, from contractors, their agents, or other parties dealing with their clients or employers in connection with work for which they are responsible.

f. When in public service as members, advisors, or employees of a governmental body or department, Engineers shall not participate in considerations or actions with respect to services provided by them or their organization in private or product engineering practice.

g. Engineers shall not solicit nor accept an engineering contract from a governmental body on which a principal, officer or employee of their organization serves as a member.

h. When, as a result of their studies, Engineers believe a project will not be successful, they shall so advise their employer or client.

i. Engineers shall treat information coming to them in the course of their assignments as confidential, and shall not use such information as a means of making personal profit if such action is adverse to the interests of their clients, their employers, or the public.
   (i.1) They will not disclose confidential information concerning the business affairs or technical processes of any present or former employer or client or bidder under evaluation, with out his consent.
   (i.2) They shall not reveal confidential information nor findings of any commission or board of which they are members.
   (i.3) When they use designs supplied to them by clients, these designs shall not be duplicated by the Engineers for others without express permissions.
   (i.4) While in the employ of others, Engineers will not enter promotional efforts or negotiations for work or make arrangements for other employment as principals or to practice in connection with specific projects for which they have gained particular and specialized knowledge without the consent of all interested parties.

j. The Engineer shall act with fairness and justice to all parties when administering a construction (or other) contract.

k. Before undertaking work for others in which Engineers may make improvements, plans, designs, inventions, or other records which may justify copyrights or patents, they shall enter into a positive agreement regarding ownership.

l. Engineers shall admit and accept their own errors when proven wrong and refrain from distorting or altering the facts to justify their decisions.

m. Engineers shall not accept professional employment outside of their regular work or interest without the knowledge of their employers.

n. Engineers shall not attempt to attract an employee from another employer by false or misleading representations.

o. Engineers shall not review the work of other engineers except with the knowledge of such Engineers, or unless the assignments/or contractual agreements for the work have been terminated.
   (o.1) Engineers in governmental, industrial or educational

employment are entitled to review and evaluate the work of other engineers when so required by their duties.

(o.2) Engineers in sales or industrial employment are entitled to make engineering comparisons of their products with products of other suppliers.

(o.3) Engineers in sales employment shall not offer nor give engineering consultation or designs or advice other than specifically applying to equipment, materials or systems being sold or offered for sale by them.

5. Engineers shall build their professional reputation on the merit of their services and shall not compete unfairly with others.

a. Engineers shall not pay nor offer to pay, either directly or indirectly, any commission, political contribution, or a gift, or other consideration in order to secure work, exclusive of securing salaried positions through employment agencies.

b. Engineers should negotiate contracts for professional services fairly and only on the basis of demonstrated competence and qualifications for the type of professional service required.

c. Engineers should negotiate a method and rate of compensation commensurate with the agreed upon scope of services. A meeting of the minds of the parties to the contract is essential to mutual confidence. The public interest requires that the cost of engineering services be fair and reasonable, but not the controlling consideration in selection of individuals or firms to provide these services.

(c.1) These principles shall be applied by Engineers in obtaining the services of other professionals.

d. Engineers shall not attempt to supplant other Engineers in a particular employment after becoming aware that definite steps have been taken toward the others' employment or after they have been employed.

(d.1) They shall not solicit employment from clients who already have Engineers under contract for the same work.

(d.2) They shall not accept employment from clients who already have Engineers for the same work not yet completed or not yet paid for unless the performance or payment requirements in the contract are being litigated or the contracted Engineers' services have been terminated in writing by either party.

(d.3) In case of termination of litigation, the prospective Engineers before accepting the assignment shall advise the Engineers being terminated or involved in litigation.

e. Engineers shall not request, propose nor accept professional commissions on a contingent basis under circumstances under which their professional judgments may be compromised, or when a contingency provision is used as a device for promoting or securing a professional commission.

f. Engineers shall not falsify nor permit misrepresentation of their, or their associates', academic or professional qualifications. They shall not misrepresent nor exaggerate their degree of responsibility in or for the subject matter of prior assignments. Brochures or other presentations incident to the solicitation of employment shall not misrepresent pertinent facts concerning employers, employees, associates, joint ventures, or their past accomplishments with the intent and purpose of enhancing their qualifications and work.

g. Engineers may advertise professional services only as a means of identification and limited to the following:

(g.1) Professional cards and listings in recognized and dignified publications, provided they are consistent in size and are in a section of the publication regularly devoted to such professional cards and listings. The information displayed must be restricted to firm name, address, telephone number, appropriate symbol, names of principal participants and the fields of practice in which the firm is qualified.

(g.2) Signs on equipment, offices and at the site of projects for which they render services, limited to firm name, address, telephone number and type of services, as appropriate.

(g.3) Brochures, business cards, letterheads and other factual representations of experience, facilities, personnel and capacity to render service, providing the same are not misleading relative to the extent of participation in the projects cited and are not indiscriminately distributed.

(g.4) Listings in the classified section of telephone directories, limited to name, address, telephone number and specialties in which the firm is qualified without resorting to special or bold type.

h. Engineers may use display advertising in recognized dignified business and professional publications, providing it is factual, and relates only to engineering, is free from ostentation, contains no laudatory expressions or implication, is not misleading with respect to the Engineers' extent of participation in the services or projects described.

i. Engineers may prepare articles for the lay or technical press which are factual, dignified and free form ostentations or laudatory implications. Such articles shall not imply other than their direct participation in the work described unless credit is given to others for their share of the work.

j. Engineers may extend permission for their names to be used in commercial advertisements, such as may be published by manufacturers, contractors, material suppliers, etc., only by means of a modest dignified notation acknowledging their participation and the scope thereof in the project or product described. Such permission shall not include public endorsement of proprietary products.

k. Engineers may advertise for recruitment of personnel in appropriate publications or by special distribution. The information presented must be displayed in a dignified manner, restricted to firm name, address, telephone number, appropriate symbol, names of principal participants, the fields of practice in which the firm is qualified and factual descriptions of positions available, qualifications required and benefits available.

l. Engineers shall not enter competitions for designs for the purpose of obtaining commissions for specific projects, unless provision is made for reasonable compensation for all designs submitted.

m. Engineers shall not maliciously or falsely, directly or indirectly, injure the professional reputation, prospects, practice or employment of another engineer, nor shall they indiscriminately criticize another's work.

n. Engineers shall not undertake nor agree to perform any engineering service on a free basis, except professional services which are advisory in nature for civic, charitable, religious or non-profit organizations. When serving as members of such organizations, engineers are entitled to utilize their personal engineering knowledge in the service of these organizations.

o. Engineers shall not use equipment, supplies, laboratory nor office facilities of their employers to carry on outside private practice without consent.

p. In case of tax-free or tax-aided facilities, engineers should not use student services at less than rates of other employees of comparable competence, including fringe benefits.

6. Engineers shall act in such a manner as to uphold and enhance the honor, integrity and dignity of the profession.

a. Engineers shall not knowingly associate with nor permit the use of their names nor firm names in business ventures by any person or firm which they know, or have reason to believe, are engaging in business or professional practices of a fraudulent or dishonest nature.

b. Engineers shall not use association with non-engineers, corporations, nor partnerships as 'cloaks' for unethical acts.

7. Engineers shall continue their professional development throughout their careers, and shall provide opportunities for the professional development of those engineers under their supervision.

a. Engineers shall encourage their engineering employees to further their education.

b. Engineers should encourage their engineering employees to become registered at the earliest possible date.

c. Engineers should encourage engineering employees to attend and present papers at professional and technical society meetings.

d. Engineers should support the professional and technical societies of their disciplines.

e. Engineers shall give proper credit for engineering work to those to whom credit is due, and recognize the proprietary interests of others. Whenever possible, they shall name the person or persons who may be responsible for designs, inventions, writings or other accomplishments.

f. Engineers shall endeavor to extend the public knowledge of engineering, and shall not participate in the dissemination of untrue, unfair or exaggerated statements regarding engineering.

g. Engineers shall uphold the principle of appropriate and adequate compensation for those engaged in engineering work.

h. Engineers should assign professional engineers duties of a nature which will utilize their full training and experience insofar as possible, and delegate lesser functions to subprofessionals or to technicians.
i. Engineers shall provide prospective engineering employees with complete information on working conditions and their proposed status of employment, and after employment shall keep them informed of any changes.

(Reprinted with permission of the Accreditation Board for Engineering and Technology, Inc. [ABET]).

Figure 2.6: IEEE CODE OF ETHICS

Preamble

Engineers, scientists and technologists affect the quality of life for all people in our complex technological society. In the pursuit of their profession, therefore, it is vital that IEEE members conduct their work in an ethical manner so that they merit the confidence of colleagues, employers, clients and the public. This IEEE Code of Ethics represents such a standard of professional conduct for IEEE members in the discharge of their responsibilities to employers, to clients, to the community and to their colleagues in this Institute and other professional societies.

Article I

Members shall maintain high standards of diligence, creativity, and productivity, and shall:

1. Accept responsibility for their actions;
2. Be honest and realistic in stating claims or estimates from available data;
3. Undertake technological tasks and accept responsibility only if qualified by training or experience, or after full disclosure to their employers or clients of pertinent qualifications;
4. Maintain their professional skills at the level of the state of the art, and recognize the importance of current events in their work;
5. Advance the integrity and prestige of the profession by practicing in a dignified manner and for adequate compensation.

Article II

Members shall, in their work:

1. Treat fairly all colleagues and co-workers, regardless of race, religion, sex, age or national origin;
2. Report, publish and disseminate freely information to others, subject to legal and proprietary restraints;
3. Encourage colleagues and co-workers to act in accord with this Code and support them when they do so;
4. Seek, accept and offer honest criticism of work, and properly credit the contributions of others;
5. Support and participate in the activities of their professional societies;
6. Assist colleagues and co-workers in their professional development.

Article III

Members shall, in their relations with employers and clients:

1. Act as faithful agents or trustees for their employers or clients in professional and business matters, provided such actions conform with other parts of this Code;
2. Keep information on the business affairs or technical processes of an employer or client in confidence while employed, and later, until such information is properly released, provided such actions conform with other parts of this Code;
3. Inform their employers, clients, professional societies or public agencies or private agencies of which they are members or to which they may make presentations, of any circumstance that could lead to a conflict of interest;
4. Neither give nor accept, directly or indirectly, any gift, payment or service of more than nominal value to or from those having business relationships with their employers or clients;
5. Assist and advise their employers or clients in anticipating the possible consequences, direct and indirect, immediate or remote, of the projects, work or plans of which they have knowledge.

Article IV

Members shall, in fulfilling their responsibilities to the community:

1. Protect the safety, health and welfare of the public and speak out against abuses in these areas affecting the public interest;
2. Contribute professional advice, as appropriate, to civic, charitable or other nonprofit organizations;
3. Seek to extend public knowledge and appreciation of the profession and its achievements.

Figure 2.7: ASME CODE OF ETHICS OF ENGINEERS

The Fundamental Principles

Engineers uphold and advance the integrity, honor, and dignity of the Engineering profession by:

I. using their knowledge and skill for the enhancement of human welfare;
II. being honest and impartial, and serving with fidelity the public, their employers and clients, and
III. striving to increase the competence and prestige of the engineering profession.

The Fundamental Canons

1. Engineers shall hold paramount the safety, health and welfare of the public in the performance of their professional duties.
2. Engineers shall perform services only in the areas of their competence.
3. Engineers shall continue their professional development throughout their careers and shall provide opportunities for the professional development of those engineers under their supervision.
4. Engineers shall act in professional matters for each employer or client as faithful agents or trustees, and shall avoid conflicts of interest.
5. Engineers shall build their professional reputation on the merit of their services and shall not compete unfairly with others.
6. Engineers shall associate only with reputable persons or organizations.
7. Engineers shall issue public statements only in an objective and truthful manner.

Figure 2.8: CODE OF ETHICS FOR U.S. GOVERNMENT SERVICE (APPENDIX II)

HOUSE CONCURRENT RESOLUTION 175, 85TH CONGRESS, 2D SESSION

Resolved by the House of Representatives (the Senate concurring), That it is the sense of the Congress that the following Code of Ethics should be adhered to by all Government employees, including officeholders:

Code of Ethics for Government Service

Any person in Government service should:

1. Put loyalty to the highest moral principles and to country above loyalty to persons, party, or Government department.
2. Uphold the Constitution, laws, and legal regulations of the United States and of all governments therein and never be a party to their evasion.
3. Give a full day's labor for a full day's pay; giving to the performance of his duties his earnest effort and best thought.
4. Seek to find and employ more efficient and economical ways of getting tasks accomplished.
5. Never discriminate unfairly by the dispensing of special favors or privileges to anyone, whether for remuneration or not; and never accept, for himself or his family, favors or benefits under circumstances which might be construed by reasonable persons as influencing the performance of his governmental duties.
6. Make no private promises of any kind binding upon the duties of office, since a Government employee has no private word which can be binding on public duty.
7. Engage in no business with the Government, either directly or indirectly, which is inconsistent with the conscientious performance of his governmental duties.
8. Never use any information coming to him confidentially in the performance of governmental duties as a means for making private profit.
9. Expose corruption wherever discovered.
10. Uphold these principles, ever conscious that public office is a public trust.

3

Differences Between Professions

The ethics of the medical, legal, and engineering professions have both similarities and differences. In fact, there are differences between the codes which define ethics in the various engineering professions. The reasons for the differences are not easy to ascertain, but an effort will be made to point some of them out. The differences between the codes of ethics of professions and the ethics of nonprofessional vocations is even greater than the differences between the professions.

One of the areas where the greatest differences occur is in the advertising profession. This is not new. Advertisers consider it to be to their advantage to portray their products as having superior and magical properties which are not only unique, but which are also capable of providing users and owners with many desirable characteristics such as allure and the ability to portray themselves as people of great taste and great social accomplishment. This appeal to the snobbishness of the public, and to the ability of a product to result in a person having greater appeal to the public and to the opposite sex, is typified in ads for clothes, jeans, and automobiles, to name a few. Of course tobacco, beer, and other liquor advertisements are of similar nature.

Advertising is a highly developed skill. The question of the differences between engineering and the advertising ethics is well expressed: "The trade of advertising is now so near to perfection that it is not easy to propose any improvement. But as every art ought to be exercised in due subordination to the public good, I cannot suppose it as moral question to those masters of the ear whether they do sometimes play too wantonly with our passions." The author of this statement is not a contemporary. It was written by Samuel Johnson who died in 1784. What was said then is just as true today.

Modern advertising is often designed to stir passions. An example is, "Merrill Lynch is bullish on America." The thundering herd which was shown in the advertisement conveys no message of the capabilities of the advertiser nor are they trying to sell cattle. The advertisement is meant to convey the concept that you too should trust your investing to the advertiser in the hope of receiving some service you desire.

The ethics of the marketplace are not very strict. Of course there are laws about what can be said on labels and what can be said in advertising, but it is quite easy to imply a great many things without saying them. Another common advertising technique is the comparison ad seen daily on television screens. Breakfast cereals are advertised this way. A cereal is touted for its content of all the needed vitamins or other daily diet essentials. The comparison is then made that shows that not one of several other cereals has near the content. It would take many bowls of the other to provide similar vitamin content. The argument that the content of breakfast cereals should be anything but vitamins is lost in the telling. In fact, some authorities have claimed that the main value of eating some dry breakfast cereals is the milk that is eaten with it.

Most professional codes of ethics infer that the practitioner not say anything ill of another professional in the group. The adherence to this practice is sometimes difficult. It is perfectly proper to disprove a theory that another professional propounds. It is, however, considered improper to brand that professional as a charlatan or an uneducated individual or to claim that he or she is devoid of any ability. This practice can be carried too far. In some professions there are incompetents practicing who either do not know the basics, or if they do, ignore the facts and the theory in the manner in which they treat subjects or clients. Thus some codes require that the members of the profession identify such malpractice for the benefit of the public and the profession.

These are some of the reasons why it is considered unethical to try and perform tasks for which you are not properly trained. There are problems here also. Who among us was properly trained to explore the problems of space before the early satellites were launched? In this instance it was common knowledge that nobody had made any product with a record of consistent performance in a space environment. There were people who were able to use available knowledge to describe the problems and analyze how certain products would perform and what characteristics were necessary to assure performance. The early satellites were tiny objects, said to be the size of grapefruit. It was easy to provide launching facilities for these small objects which do not begin to compare in complexity with those required today to launch vehicles carrying astronauts and space loads. The power to operate the tiny radio transmitters carried into space was also state of the art.

The work of many engineers involved in the space endeavor resulted in the launching of small satellites which were supposed to be shut off in a few months. At least one of these had two failures. The first was the failure of the shutdown switch to operate before the predicted end of life of the battery power source. The second was the failure to predict the useful battery life. The battery continued to operate for a long time after the predicted end of life. This resulted in the little chatterbox continuing to send signals long after they were desired and predicted.

Fortunately these failures were not nearly as serious as a failure to properly design a structure, to provide electrical insulation and isolation, or to provide products free from chemical or biological content that is detrimental to human, animal, or plant life.

The codes of ethics of most of the professions are aimed at protecting life and the environment. Generally, in engineering codes, there are statements that engineers shall protect the public safety. Most localities have enacted laws requiring building and structures erected in the environs covered by the statute to bear the signature and sometimes the seal, of a licensed engineer or architect. To obtain this license persons must have been able to demonstrate their competence to a licensing board. According to law they are then fully qualified to design and attest to the safety of structures. The first punishment expected for failing to observe the legal requirements would be loss of license. This may require a more severe infraction than just ethical misbehavior. It would probably occur when there is a failure of a structure and the loss of property and/or injury to one or more individuals.

There are unfortunately examples of these failures: the collapse of the roofs of the Hartford Connecticut Civic Center, the C. W. Post gymnasium; the collapse of the Kansas City Grand Hyatt Skywalk; the collapse of the Tacoma Narrows Bridge; of the Mianus River Bridge on Interstate 95; and the collapse of the L'Ambiance Plaza in Bridgeport, Connecticut, during construction with the loss of 28 lives.

Are these criminal actions or ethical failures? In some instance they are both. It would be wise to review these cases and for each of us to decide for himself or herself whether they represent ethical failures. It would seem that since these events were related to structures, the code of the American Society of Civil Engineers ought to be used.

When we discuss building failures it is interesting to realize that structures have lives. A structure cannot be expected to continue to serve its original purpose without maintenance procedures being performed periodically. There is little point to criticize design, construction, or a builder when a structure has been in service for 100 years. How long should a building last? The Statue of Liberty and the Eiffel Tower have each been in place for over 100 years. Each has required major reconditioning for safety and appearance. The Brooklyn Bridge is also more than 100 years old. A recent review of its condition found the masonry towers in excellent condition. The Williamsburg Bridge which spans the East River, as does the Brooklyn Bridge, has many more problems. It was built without many of the protective features that are present on the Brooklyn Bridge. It will cost much more to recondition despite the fact that it is not as old.

How long ought a bridge last? Thousands of roadway bridges have been torn down and replaced with grander bridges as the highways have improved. None-theless there are many old bridges. All should have regular inspections. Some of the bridges need maintenance and others ought to be closed and replaced.

The length of time a structure or device is expected to last is termed life expectancy. How long should designers and builders be responsible for bridges, assuming they were not responsible for upkeep? How long should manufacturers be responsible for their products? In law, statutes indicating the period of responsibility are statutes of repose. If someone were injured by a product or structure would the maker have a responsibility to compensate the victim if the accident happened a year, 10 years, 25 years, or 100 years after construction? The L'Ambiance Plaza accident happened during construction. The Mianus River Bridge had been in place for approximately 20 years. Inspections were made during its life, but the inspectors did not look at the crucial support.

The U. S. Congress in its consideration of product liability legislation which it has yet to pass, has selected a 25-year period of repose for buildings in some of its proposals. This does not negate the fact that owners and maintenance personnel may have continuing responsibilities.

What would seem to you to be a reasonable statute of repose for various buildings and structures, tools, elevators, automobiles, plumbing systems, electrical systems, heating systems, roads, bridges, highways, office buildings, factories, and apartment buildings?

In Chapter 2 we discussed portions of the codes of ethics of engineering and other societies. The Code of Ethics of the ASCE covers civil engineers. It is closely patterned after the ABET code. We ought to go back to the essence of the chapter, the difference between some of the professions. Engineers deal mainly with products and processes. Lawyers deal with people, their freedom, their property, their rights, and their claims. Doctors deal primarily with the health and well-being of individuals. The dichotomy between what engineers do and what some other professionals do is not absolute.

Some industrial engineers deal with action, motion, and time studies and the use and scheduling of people in the industrial enterprise system, in government, and in other endeavors. Engineers are interested in the well-being of people. When employed in the scheduling and task-setting phase of the economy, engineers are working for employers and their first interests are in efficiency, economy, and getting the most out of employees. This is essentially the efficient use of manpower. On the opposite side of such efforts are similar engineers employed by some of the major union organizations seeking to protect individual members from abuse and exploitation. In some industries the union does the time study and the scheduling for small companies. In these cases it is hard to know whether the main interest of engineers is for employers or employees. Perhaps there is a middle ground and a middle interest.

Engineers employed in these applications become, or at least have an opportunity to become, more people rather than product oriented. In fact, these same engineers may become imbued with the concepts and teachings espoused by the early industrial engineer, Frederick W. Taylor, and become so interested in scheduling and output quantity, that they lose sight of the importance of doing the job correctly and obtaining conforming and quality product. It is just this phase of the industrial engineering activity that can be subject to abuse and can overwork people to the detriment of employees, products, and employers.

Lawyers, as noted, are primarily interested in people, but these same professionals, under different conditions can become product and process centered. These conditions obtain when lawyers are involved with regulations relating to processing material for a pharmaceutical or surgical organization. Compliance testing and conformance of the product to laws and regulations, and how this is determined and substantiated, become legal as well as scientific matters. Another situation that involves lawyers with products and their performance is contractual negotiations and the settlement of contract disputes. When physicians or lawyers become concerned primarily with a corporate entity that makes a product or controls a process or is involved with patent litigation, it is quite possible that these professionals become as interested and involved in making product and process decisions and defining limits and differences as engineers.

Engineers are concerned with safety, or at least are enjoined to consider the safety of the public in designing, constructing, distributing, and providing controls for products, buildings, or processes. In this type of endeavor engineers become involved, almost as lawyers do in developing standards and codes for safety and conformance. There are many types of standards, but most can be classified into several general forms. One broad general category includes standards that specify shape, form, size, tolerance, performance, and test procedures. These are sometimes described as standards that provide SMMP; that is a standard method of measuring performance and compliance. Another category is exemplified by standards that describe procedures such as test methods, sampling tables, and other procedures to achieve an end. Some standards have both parts under one cover.

Engineers may be involved in writing standards that are classified as codes. These include the ASME Boiler Code, the IEEE National Electrical Safety Code, the National Fire Protection Associations's National Electrical Code, and others.

They are so well developed and have existed for so many years, that they are recognized as necessary safety codes and are adopted into federal, state, and local laws. Some areas that do not adopt the code adopt a strengthened or modified version of the code developed under their own aegis.

The safety standards of the Underwriters Laboratories and the Standards of the Society of Automotive Engineers are widely observed, though there may not be laws that prohibit the sale and distribution of products that do not conform. Many localities have building codes that specify compliance with Underwriters' standards. Also permitted are listings by an accredited testing laboratory, which are necessities for the approved use of any product. In writing and administering these standards, there is an interface between users and producers which is more people oriented and requires both an appreciation of human relationships and physical laws.

In 1982 a lawsuit was initiated against the American National Standards Institute, Inc. (ANSI) and the American Society of Mechanical Engineers, Inc. (ASME) by the Hydrolevel Corporation. Hydrolevel claimed that the ASME and ANSI, through their position as interpreters of a code written by ASME and promulgated as an American National Standard with an ANSI/ASME identifier and number, had wrongfully excluded from the market a patented device designed and developed by Hydrolevel.

The interpreters of the code implied that the Hydrolevel device was unsafe and unsuitable for installation and that it did not comply with the code. The interpreters, acting for ASME, were two individuals who were volunteers and not employed by ASME. They worked for organizations with financial interests in a competitive device which stood to lose some business should the Hydrolevel invention be introduced.

It was claimed that ASME and ANSI had failed to properly supervise and control those who represented them and had failed to set up rules which should have properly controlled activities and prevented such action on the part of an interested interpreter.

ANSI settled out of court for approximately $50,000 leaving ASME to defend or settle the suit on its own. It is not uncommon for two defendants to view the value of a civil suit differently and to take such action as each deems most advisable to protect its interests.

The ASME lost the suit and was assessed some $4.5 million. You could question the ethics of the individuals who rendered the opinion. I do not know what action was taken against them. As a result of this case many societies took on the task of reviewing their practices and of seeing that they were not placed in a similar position. ANSI also revised its practices.

Some societies have expanded their interest in the problem of ethics beyond the position of promulgating a code. The IEEE thinks there is an ethical position the society should take when its members or other engineering professionals are faced with loss of income and other penalties for acting in the public interest. This requires that the society take a stance on many public interests. For this purpose the IEEE created a Committee on the Social Implications of Technology.

When someone receives what he or she believes is improper medical service which results in physical, mental, economic, or other loss, he or she is apt to sue the doctor or medical services provider. This is particularly true in our litigious society. Doctors have been sued for malpractice resulting in a major or minor award or no recompense. As with other professions some small percentage of the practitioners are incompetent, unethical, or otherwise unable to render reliable services. These individuals seem to be responsible for more malpractice suits than others in the same specialty. In addition some specialties suffer greater malpractice losses than others. There are also differences among cities and between urban and rural areas. Premiums for malpractice insurance can run up to $20,000.

It is not surprising that the American Medical Association (AMA), the largest medical professional society in the United States, has been engaged in trying to properly define medical ethics. There is a code of ethics for physicians and a 50-page booklet, *Current Opinions on Ethical and Judicial Affairs.*

The AMA's Principles of Medical Ethics appear in Chapter 2. It will be noted that this code also requires the continuing education of physicians and their responsibilities to society, to other health professionals, to their patients, and to themselves. These responsibilities seem to be common to many, but not all, professional organizations.

The American Society of Civil Engineers (ASCE) and the IEEE have, in developing their codes of ethics, followed the model of ABET, although the canons or principles may not be stated in the manner suggested by ABET. A review of the similarities and differences may cause you to wonder what reasoning was involved in adopting a code.

The first part of the code promulgated by ABET is termed fundamental principles. "Engineers uphold and advance the integrity, honor and dignity of the engineering profession by using their knowledge and skill for the enhancement of human welfare" [12]. The ASME and the ASCE have adopted this statement into their code. The IEEE does not specifically quote this phrase, and the ASQC, whose code least resembles the suggested code, does not adopt this phrase specifically. Why would the suggested phrase be more agreeable to the ASME and the ASCE than the other societies? Is this a difference in the duties and jobs of the individuals in the societies or could it be an option that the committee writing the code felt was more acceptable to their management and/or their members?

The second fundamental principle of the ABET Code of Ethics for engineers states: "Engineers uphold the integrity, honor and dignity of the engineering profession by: being honest and impartial, and serving with fidelity the public, their employers and clients" [13]. This appears as the second principle in the code of ethics of the ASCE. It also appears in the ASME code. It does not appear as a separate principle in the code of the IEEE. The ASQC code adopts this statement as the first principle but changes the word *fidelity* to *devotion*. The reason for this being the first rather than the second fundamental principle is not quite clear. Why did the writers replace *fidelity* with *devotion*? The differences between these definitions is worth discussing. The two words are synonyms but there is a difference in the meaning of the words. What do you suspect was the intent of the writers of the codes?

The third fundamental principle of the ABET code enjoins engineers to strive to increase the competence and prestige of the engineering profession. The ASCE and the ASME both cite this principle in their codes. The ASQC adopted the same wording. The IEEE adopts similar goals in various portions of its code, but adds other conditions. Article I clause 5 of the IEEE Code of Ethics states that "Members shall ... advance the integrity and prestige of the profession by practicing in a dignified manner and for adequate compensation" [14]. In Article II clause 6 the engineer is admonished to assist colleagues and co-workers in their professional development, and in Article IV clause 3 the member is required to "seek to extend public knowledge and appreciation of the profession and its achievements." One wonders whether all of these requirements sum up to be equivalent to the fundamental principle of the ABET code or whether the writers of the IEEE code had greater or lesser requirements than the writers of the other codes.

The code of ethics of the legal profession is complex. As noted in Chapter 2 the American Bar Association has two publications: the *Model Code of Professional Responsibility and Code of Judicial Conduct* adopted in 1969 and revised periodically up until August 1980; and the *Model Rules of Professional Conduct and Code of Judicial Conduct* adopted in August 1984. The rules must be adopted by individual states, and as of this writing many states have not adopted the newer code and rules. Additionally, a state may have other examples, rules, and regulations. The rules include references to competence, diligence, confidentiality of information, conflict of interest, successive government and private employment, professional misconduct, and reporting professional misconduct. There are many other facets of the rules which we will not discuss now.

In many ways the sections that have been mentioned cover activities, situations, and questions that seem to be similar in the engineering, quality control, and legal professions. This can be particularly true for those who are top-level quality managers and for those who are in the engineering and quality profession and are acting as consultants working with an organization. For those of us doing design and testing there are fewer resemblances.

The question of competence is one which members of all professions find in their code of ethics. Just what is competence? The legal profession in the *Model*

Rules points out that a member newly appointed to the bar or a newly trained lawyer may have as much or more expertise for a specific case than a lawyer with a lot of experience. The reverse may also be true in other instances. Where lawyers lack the expertise, it is proper for them to obtain associates who have the expertise and have them do much of the work on the case. In the engineering profession the same practice is pursued. When engineers and quality professionals are employees of a corporation, the situation can become quite difficult. Employees may not feel that they can tell superiors that they do not have the expertise, the knowledge, or the experience to handle the task. What are the ethics of such a situation? There is no restriction in any code that prevents a professional from researching the literature and getting the required information. Would it be proper to say that a professional who is able to rapidly obtain the information on the subject so that it can be handled expeditiously is in possession of the required expertise?

In the legal profession where an incident has occurred which resulted in litigation, it is not unusual that the attorney requires help from experts to obtain information in the fields of medicine, economics, science, engineering, or some other technology. In such cases the practice is to find an individual, with requisite experience and expertise, who is familiar with the specific field. There are instances where the judge also needs such advice and may hire an individual to advise the court on the relative merits of the claims of opposing experts or be the expert for the court in the event there are no other experts. There are instances where lawyers have the ability to research the literature and consult with experts in the field to the extent that they themselves may know more about a particular injury and its effect on a person than almost anyone else. These lawyers still would not consider themselves experts and would never try to treat anyone suffering from this injury or disease.

As might be suspected not all experts are equal. The jury may not believe the expert who presents an opinion based on intimate knowledge of the situation and may believe a less qualified individual who is glib and whose presentation is in reality a complete fabrication. The problem is that when the presentations are made in court there may be no way to really rebut and challenge some opinions. To make the situation a little more controllable some states make it necessary for the two sides in a litigation to identify their experts well before trial and to allow these experts to be deposed before trial. A deposition is a questioning under oath which can be read into the trial. It provides a method whereby the opinions are available for review before trial and can be challenged if in error. Another procedure is that of requiring a written report. This statement of facts and opinions before trial can be made available to opposing sides. A skillful expert can present a well-documented and sustainable case. In other instances a skillful expert can comment on a report that is all bluster and not very factual. Some reports lack real substance despite having been produced by a professional. This may be because the professional lacks the required skill and expertise in the field, even though well skilled in other fields. Reports of this nature have also been produced by prestigious engineers and others who are trying to help their side win a case. Whether the attorney who has commissioned the report and the client are aware of this deception is not usually known to the opponent. Is the production of such a report ethical or are there special conditions under which such a report is likely to be ethical?

There are other situations where attorneys and judges have been misled by the professed knowledge of someone they believe to be an expert. In one case a judge's opinion in a patent litigation was written by an individual who professed to be an expert on the subject. It was patently absurd and the decision was later overturned at considerable expense. The judge was misled. Did the judge's expert act ethically? Did he really believe that he was an expert or was he too ignorant to understand any of the facts and theories relating to the matter? Does ignorance excuse or provide immunity against an unethical act?

In another case the judge had the expertise required to settle a question. The plaintiff's expert claimed that a large ring had failed and caused the accident. The failure was due to the fact that the ring, which surrounded a large disc, had an internal diameter smaller than the disc to which it was bolted. The defense objected and said that this statement was ridiculous. The expert claimed that the drawings of the defendant specified this situation. The judge said he just happened to be a registered Professional Engineer in addition to being an attorney. He asked to see the drawings and after examining them agreed with the plaintiff's expert. This was a simple situation and an elegant solution. The most probable cause of the errors on the drawings is that either there was some incompetence among the personnel or a faulty system. Incompetence in industry among people who have been trained in the fundamentals is really not easily explained. In industry there are people posing as engineers and quality professionals who have had almost no training. In many of these instances the supervisors of these industrial employees are even more ignorant than the employees. The problems that arise in industry because of these incompetencies is legion. They are best explained by Parkinson's Laws.

The University of the State of New York's State Education Department published the *Professional Engineering Handbook* in December 1988. Of particular interest are the Rules of the Board of Regents on Unprofessional Conduct. These are included in the following figure showing the contents of Section 29. Many of the restrictions apply to other professions as well as engineering and land surveying.

Figure 3.1: STATE OF NEW YORK RULES OF THE BOARD OF REGENTS ON UNPROFESSIONAL CONDUCT

Section 29.1

GENERAL PROVISIONS FOR ALL PROFESSIONS

(a) Unprofessional conduct shall be the conduct prohibited by this section. The provisions of these rules applicable to a particular profession may define additional acts or omissions as unprofessional conduct and may establish exceptions to these general prohibitions.

(b) Unprofessional conduct in the practice of any profession licensed or certified pursuant to title VIII of the Education Law shall include:

(1) willful or grossly negligent failure to comply with substantial provisions of Federal, State or local laws, rules or regulations governing the practice of the profession;

(2) exercising undue influence on the patient or client, including the promotion of the sale of services, appliances or drugs in such manner as to exploit the patient or client for the financial gain of the practitioner or of a third party;

(3) directly or indirectly offering, giving, soliciting, or receiving or agreeing to receive, any fee or other consideration to or from a third party for the referral of a patient or client or in connection with the performance of professional services;

(4) permitting any person to share in the fees for professional services, other than: a partner, employee, associate in a professional firm or corporation, professional subcontractor or consultant authorized to practice the same profession, or a legally authorized trainee practicing under the supervision of a licensed practitioner. This prohibition shall include any arrangement or agreement whereby the amount received in payment for furnishing space, facilities, equipment or personnel services used by a professional licensee constitutes a percentage of, or is otherwise dependent upon, the income or receipts of the licensee from such practice, except as otherwise provided by law with respect to a facility licensed pursuant to article 28 of the Public Health Law or article 13 of the Mental Hygiene Law;

(5) conduct in the practice of a profession which evidences moral unfitness to practice the profession;

(6) willfully making or filing a false report, or failing to file a report required by law or by the Education Department, or willfully impeding or obstructing such filing, or inducing another person to do so;

(7) failing to make available to a patient or client, upon request, cop-

ies of documents in the possession or under the control of the licensee which have been prepared for and paid for by the patient or client;

(8) revealing of personally identifiable facts, data or information obtained in a professional capacity without the prior consent of the patient or client, except as authorized or required by law;

(9) practicing or offering to practice beyond the scope permitted by law, or accepting and performing professional responsibilities which the licensee knows or has reason to know that he or she is not competent to perform, or performing without adequate supervision professional services which the licensee is authorized to perform only under the supervision of a licensed professional, except in an emergency situation where a person's life or health is in danger;

(10) delegating professional responsibilities to a person when the licensee delegating such responsibilities knows or has reason to know that such person is not qualified, by training, by experience or by licensure, to perform them;

(11) performing professional services which have not been duly authorized by the patient or client or his or her legal representative;

(12) advertising or soliciting for patronage that is not in the public interest:

   (i) Advertising or soliciting not in the public interest shall include but not be limited to advertising or soliciting that:

      (a) false, fraudulent, deceptive, misleading, sensational or flamboyant;
      (b) represents intimidation or undue pressure
      (c) uses testimonials;
      (d) guarantees any service;
      (e) makes any claim relating to professional services or products or the cost or price therefor which cannot be substantiated by the licensee, who shall have the burden of proof;
      (f) makes claims of professional superiority which cannot be substantiated by the licensee, who shall have the burden of proof;
      (g) offers bonuses or inducements in any form other than a discount or reduction in an established fee or price for a professional service or product;

   (ii) The following shall be deemed appropriate means of informing the public of the availability of professional services:

      (a) informational advertising not contrary to the foregoing prohibitions; and

(b) the advertising in a newspaper, periodical or professional directory or on radio or television of fixed prices, or a stated range of prices, for specified routine professional services, provided that there is an additional charge for related services which are an integral part of the overall service being provided by the licensee and the advertisement shall so state, and provided further that the advertisement indicates the period of time for which the advertised prices shall be in effect:

(iii) (a) all licensees placing advertisements shall maintain, or cause to be maintained, an exact copy of each advertisement, transcript, or videotape thereof as appropriate for the medium used, for a period of one year after its last appearance. This copy shall be made available for inspection upon demand of the Education Department or in the case of physicians, physicians's and specialist's assistants, the Department of Health;

(b) a licensee shall not compensate or give anything of value to representatives of the press, radio, television or other communications media in anticipation of or in return for professional publicity in a news item;

(iv) No demonstrations, dramatizations or other portrayals of professional practice shall be permitted in advertising on radio or television;

(13) failing to respond within 30 days to written communications from the Education Department or the Department of Health and to make available any relevant records with respect to an inquiry or complaint about the licensee's unprofessional conduct. The period of 30 days shall commence on the date when such communication was delivered personally to the licensee. If the communication is sent from either department by registered or certified mail, with return receipt requested, to the address appearing in the last registration the period of 30 days shall commence on the date of delivery to the licensee, as indicated by the return receipt.

(14) violating any term of probation or condition or limitation imposed on the licensee by the Board of Regents pursuant to Education Law section 6511.

(From *Professional Engineering Handbook* [New York: The University of the State of New York, The State Education Dept.], Dec 1988.)

Figure 3.2 STATE OF NEW YORK RULES OF THE BOARD OF REGENTS ON UNPROFESSIONAL CONDUCT GENERAL PROVISIONS FOR DESIGN PROFESSIONS

Section 29.3

(a) Unprofessional conduct shall also include, in the professions of architecture and landscape architecture, engineering and land surveying:

(1) being associated in a professional capacity with any project or practice known to the licensee to be fraudulent or dishonest in character, or not reporting knowledge of such fraudulence or dishonesty to the Education Department;

(2) failing to report in writing to the owner or to the owner's designated agent any unauthorized or improperly authorized substantial disregard by any contractor of plans or specifications for construction or fabrication, when professional observation or supervision of the work is provided for in the agreement between the owner and the design professional or when supervision of the work is under the control of the design professional;

(3) certifying by affixing the licensee's signature and seal to documents for which the professional services have not been performed by, or thoroughly reviewed by, the licensee; or failing to prepare and retain a written evaluation of the professional services represented by such documents in accordance with the following requirements:

(i) A licensee who signs and seals documents not prepared by the licensee or by an employee under the licensee's direct supervision shall prepare, and retain for a period of not less than six years, a thorough written evaluation of the professional services represented by the documents, including but not limited to drawings, specifications, reports, design calculations and references to applicable codes and standards. Such written evaluation shall clearly identify the project and the documents to which it relates, the sources of the documents and the name of the person or organization for which the written evaluation was conducted, and the date of the evaluation, and the seal and signature of the licensee shall also be affixed thereto.

(ii) Nothing in this paragraph shall be construed as authorizing the practice of a design profession in this State by persons other than those authorized to practice pursuant to the provisions of Articles 145, 147 or 148 of the Education Law;

(4) Failure by a licensee to maintain for at least six years all preliminary and final plans, documents, computations, records and professional evaluations prepared by the licensee, or the licensee's employees, relating to work to which the licensee has affixed his seal and signature.

(5) have a substantial financial interest, without the knowledge and approval of the client or employer, in any product or in the bids or earnings or any contractor, manufacturer or supplier on work for which the professional has responsibility;

(6) permitting any person to share in the fees for professional services, other than: a partner, employee, associate in a professional firm or corporation, sub contractor or consultant. This prohibition shall include any arrangement or agreement whereby the amount received in payment for furnishing space, facilities, equipment, or personnel services used by a professional licensee constitutes a percentage of or is otherwise dependent upon the income or receipts of the licensee from such practice. This provision shall apply in lieu of section 29.1(b)(4) of the Part;

(7) accepting any form of compensation from more than one party for services on the same project without fully disclosing the circumstances and receiving approval from all interested parties; or

(8) participating as a member, advisor or employee or a government body in those actions or deliberations which pertain to services provided by the practitioner or his or her organization for such government body.

(b) Unprofessional conduct shall not be construed to include the employment, with the knowledge of the client, of qualified consultants to perform work in which the consultant has special expertise. This provision shall apply in conjunction with section 29.1(b)(9) of this Part.

(For a complete copy of the "Rules of the New York State Board of Regents Relating to Definitions of Unprofessional Conduct" contact the Customer Service Unit at (518) 474–3817 or 1–800–342–3729. From *Professional Engineering Handbook* [New York: The University of the State of New York, The State Education Dept.], Dec. 1988.)

4
Ethical Questions in Product and Services Industries

Many engineers and quality personnel are employed in organizations which are engaged in endeavors relating to the manufacture and delivery of product or the delivery of service. In the product-related industries there are often many layers of management, and there is a variety of endeavors ranging from the design and development of product, through the stages of procurement, manufacturing, testing, packaging, shipping, sale, installation, and service. Service industries include those to maintain a product or the delivery of service, such as electricity, maintenance, washing, or medical.

The engineer in any of these activities is often faced with ethical questions, some minor, some major, and some that are of career-determining nature. As an example let it be assumed that a product is in the design phase. Joe, an engineer, is a member of a team and is working on one part of the task. Frances, an associate is working on another set of components and is experiencing difficulties. She cannot achieve the specified performance. She realizes that Joe will recognize this when he tries to operate the two sections. Joe might also recognize that the system as it now exists is unstable and unsafe, due to Frances' portion of the task. Frances suggests to Joe that she is certain that she will correct these deficiencies with a better arrangement within a few days. She asks Joe not to report the deficiencies to the supervisor. Does Joe become deeply involved if he agrees to the suggestion? Should he agree, try to convince Frances to advise the supervisor, or should he tell the supervisor that a problem exists? Could Joe's action in this instance affect his career?

The decisions of engineers engaged in product design and development can affect the safety of the product and the safety of those manufacturing and servicing the product. Their decisions can also affect the economy of operation and in some instances the national economy.

The efficient operation of U.S. automobiles and home air conditioning and heating devices all have an effect on the total energy in the form of petroleum products consumed and therefore, on the amount of oil imported into the United States. The more efficient these devices are, the less total fuel consumed and imported. Since these devices last for many years the increase of efficiency can be beneficial for years after the product is sold. It has been said that the failure of U.S. auto manufacturers to meet the projected gasoline consumption limits which had been established for the 1988 model year, and the government's agreement to let the limits move out another year, has resulted in the use of 400,000 barrels more of crude oil per day in this country. Assuming that the more oil we use the more we must import, and that this has a negative impact on our already poor negative trade balance, was the decision an ethical one? The decision may have been at the design level, the management level, or even an unrecorded agreement between government and some industry executives. There are many facets to such an effort.

In some areas of the United States there is a water shortage. We use more water than many other countries on a per capita basis. Perhaps the largest waster is the flushing of toilets. Toilets have been developed and are available in some areas that use much less water. Is it ethical not to insist that all new installations and replacements use these more efficient toilets?

In some instances there is a widespread knowledge of the proper things to do and in others there is abysmal ignorance. If you were involved in a program dedicated to the development of a product for sale to the government or to consumers and there was a set of specifications that had to be met, then you could state that the specifications were well defined and that the developing group had no excuse for stopping before the specifications had been exceeded or met. If the specifications had not been met and the development group assured the company that they had, then the developers might be guilty of unethical action. If a large investment in tool, equipment, and parts had been made, someone might try to convince the government to accept the product, and might be successful. On the other hand this might be a financial disaster. There have been situations where this has occurred. The plant was built, and the product could not be produced or sold. There have been other instances where the specifications were not met and the data presented was fraudulent. In fact there are detailed records where the product was shipped to the government and others, even though it did not meet specifications. This can happen when the lowest engineers on the totem pole err or cheat on their data and the supervisors do not check the results. Each in turn compounds the company's commitment, resulting in the company getting itself so deeply involved that they are committed to manufacture and deliver or go bankrupt. There is then pressure on executives to cut deals and knowingly or unknowingly get the company in even deeper. What can you do as the manager of a division when there is nothing else to deliver but the product that does not really meet the specification?

I managed a plant producing nickel cadmium batteries. The plant had been built by the Corps of Engineers to produce batteries for the military. The effluent of the plant went into the town sewage system and from there it went into the Hudson River with no treatment. One might question whether the engineers of the company or the government should have been ignorant of the effect of dumping the effluent into the Hudson. Nonetheless it was done without question. Then some years later the town was advised that it must cease dumping raw sewage into the river. This meant the creation of a sewage treatment plant. We used as much water as the town, and we had our own well. We had found that the town could not supply us with water. Now the town said that it could not take our effluent. They were obligated to take our sewage but not our industrial waste. We therefore redirected the effluent through a bypass that went directly to the river. We considered that a good solution. We had changed nothing. The effluent entering the river now was no different than before, and it contained no sewage so there was no problem. It was, I can assure you, strictly ignorance. One day a federal waterways inspector knocked on the door of the plant and wanted to see our permit to discharge effluent into the river. We should have requested permission and obtained a permit. We would have to stop. We could continue only if the discharge met specific standards.

If we were to meet these water standards there was much work to be done and large installations to be made. I was assured by the chief executive that I should not worry about it. They would forestall any closure, and when we were pressed they would engage competent legal council to delay the enforcement. We had been served a notice but had made no agreement.

I was able to obtain funds to develop procedures and purchase equipment. When the time came to close down the discharge to the river those who had promised to defend, delay, and protect the plant promptly forgot their promises. We were able to run because equipment was in place. Was it ethical to abandon the concept of delay? Was it ethical to promise aid and assistance in creating delay? Part of the cadmium deposits in the Hudson River are due to these discharges. Was it proper to have discharged any of these materials into the Hudson, at any time? There are lots of ethical questions here. Is ignorance an acceptable excuse?

Engineers working for product manufacturers may be employed in the design and construction of equipment designed especially for the manufacture and assembly of a product. They would be involved in the safe control of the equipment so that the operators were not exposed to needless chance of injury. The term *needless* is chosen purposely. When machinery is employed it is often impossible to so design, construct, and operate the equipment in such a manner that is impossible to cause injury. Good examples are the ordinary sharp knife and the power saw. The knife is designed to cut and the sharper it is the easier it is to cut and control. Nonetheless a knife can do grievous injury to the careless user. The power saw is also designed to cut. It is possible to provide guards, but there must be a way of feeding work to the saw. It is possible to insert a finger in place of the work. This is likely to result in serious injury. However the use of good guards, controls, and fixtures in the work place plus other design considerations can reduce the chance of injury. These same guards can make the work rate slower and thereby create conditions under which pieceworkers and/or supervisors bypass the guards to increase the rate of output. Engineers providing good setups and well-designed guards can contribute to the safety of the work place.

One product of a meat packing plant was a compressed sausage-shaped bulk of chopped meat. It could be sliced into patties and used for hamburgers. The block of meat was first compressed into a cylinder by a clamshell and then the ends of the cylinder were compressed by hydraulic plungers coming in from each end. The machine had two hoods, one on the left and the other on the right. These hoods prevented the juices from squirting out on the operator and each carried a device which triggered a control mechanism. When both hoods were closed the machine was supposed to start. One day the machine was moved and was not level. The hoods would then slide and make it necessary for both to be kept closed during the operation. The operator complained but was told to be careful and to continue to use the machine. The foreman was charged with producing so much product that day, and he really needed the output of the machine. During one cycle, as the operator was loading the meat into the clamshell with his left hand, the right hand hood slid closed and the machine began its cycle mangling several of his fingers.

The machine should not have started with either the left or the right hood open. It did start because of a short in one of the solid-state relays created by an error in the assembly of the unit at the factory. Further analysis of the operating cycle disclosed that the safety devices were not connected the way that the manufacturer had intended. This error had been made in the factory. Had the safety devices been connected in the manner shown in the wiring diagram and described in the instruction manual, the machine would not have started even with the shorted relay.

The errors made in the factory would have been detected had the testing procedure been different. There were ethical failures. It was suggested that the production worker should have refused to operate the machine when he knew that it was not operating properly. Was it ethical to expect that he would not operate the machine until it was fixed? Was it ethical to request him to operate the machine?

Was it ethical to put the pressure on his foreman so that he felt the necessity of having the machine operate whether it was working properly or not? Was the factory ethical in providing a safety mechanism that could be faulted by a simple short-circuited relay? Was it ethical to test a device in a manner that would permit it to be shipped even though it had an undetected fault, one which would have been disclosed by another testing means? Is it ethical to let each manufacturer, large or small, set his or her own standards of what is a suitable safety mechanism and control? Can you think of other ethical questions?

Materials used in the product and the machines used in manufacture can either contribute to or may jeopardize workers' safety. The use of heavy metals such as lead, cadmium, and mercury can provide hazardous atmosphere. Other materials are also hazardous to the health. The National Institute of Occupational Safety and Health (NIOSH) has a long list of hazardous industrial materials.

Service industries also have problems with the safety of users and workers. Ethically, engineers have responsibilities to protect plant workers and users. Sometimes all of the safety efforts cannot be built into the product. This is particularly true of some service industries, for example utilities which deliver gas and electricity. The utilities distribute natural gas which has no odor, but add trace materials which have odors in order that a leak may be detected. The engineer's contribution is the various codes which govern what is considered to be safe installations for homes, offices, and factories. Engineers and other practitioners can, by participating in the writing and revision of these codes, contribute to the safety of the public. Safe practice procedures are essential. All those employed in operating jobs in these industries must be taught to observe specific practices while working in areas where they are exposed to hazards.

There are licensed installation people and inspectors who are responsible for checking the quality of workmanship and compliance with codes. Therefore, it is reasonable to expect that compliance is the usual situation. However there are changes and additions by the unlicensed, the ignorant, and the careless which provide for dangerous and accident-producing installations. Ethically is it proper to let the local handyman or friend add to the house, install gas, electric, or fuel oil systems? Is it ethical for a worker to leave an unfinished job in a hazardous condition?

I worked for a plant that produced an electronic gun used to project electrons onto the face of a television picture tube. These tubes operated at very high voltages, in the order of 25,000 volts. The current required was about one milliampere. Since these devices were manufactured, it was necessary to test them and to subject them to life tests. This meant that each device tested had to run for many hours. To supply the high voltage and control it accurately, each test socket had its own power supply. One test equipment engineer noted that these were expensive and that a considerable sum could be saved if we installed a high-voltage generator. This not only had sufficient voltage, but it could also supply several amperes and thus supply thousands of sockets although we really only needed several hundred. The cost savings were impressive.

The technician put through a requisition to buy the generator, and the plant manager approved it. I disapproved it. The manager complained up the line and in due course, I was visited by the chief executive and accused of being unwilling to save money. I explained that our technicians were not accustomed to working on hazardous lines and anyone touching any outlet or wire connected to this generator could suffer a lethal shock. The CEO still was quite insistent about saving money. I added that I had no interest in sending one of our people home in a basket, and that if he ordered the hazardous equipment installed I was going to put a letter in the corporate records. I never heard any more about the suggestion, and the high-voltage, high-current generator was not purchased. Was I being ethical? Were the plant manager and CEO being ethical or might they have been ignorant?

Other problems occur when products must be serviced in the field. Sometimes it is almost impossible to perform a minor service without moving and disassembling major parts. In some automobiles it was impossible to replace spark plugs without moving major parts of the vehicle. This not only made for a more expensive job, but it also exposed field workers to the possibility of major injuries. The workers also became responsible for doing major damage to vehicles. Is the design of such a vehicle ethical? Similarly, is the design of a device or product ethical if it is planned so that some small replacement parts are not available? When it is necessary to replace a small or insignificant part, a major assembly must be purchased and replaced. There may or may not be good reasons for this, but if the only reason is to make a large sale, is this practice ethical?

A power saw replacement part was found to be far more expensive than I had anticipated. The replacement portion was furnished, and I found that it included two new guards that brought the saw up to the same state of development as the company's newest mode. The elimination of smaller cheaper parts made it possible for the safety of the product to be greatly improved and the ease of operation was also improved. There was no recall or notification of the availability of the upgrade to owners. Was the company ethical in removing the smaller parts from its supply list? Was it ethical to make the buyer of a replacement part also purchase new guards? Should the improved upgrade kit have been provided earlier? What would have been the most ethical procedure to follow when the new design was developed?

Engineers are often managers and sometimes they are officers of corporations. In these positions engineers are responsible for many activities within corporations and thus may have ethical responsibilities far exceeding that of engineers in lesser positions. As major decision makers, the engineers in corporate management may be responsible for the well-being of employees and the public which uses the product. The law recognizes the responsibility of top-level management for some infractions of the law which may occur. When a prominent baby food supplier sold containers of apple juice which contained no apple juice but only colored and sweetened water, the law imposed penalties on the executives of the corporation. There have been other cases in which the chief executive of a firm was sentenced for violations of the law committed by the organization. (More on this in Chapter 13.)

There may be ethical or moral responsibilities which engineers recognize. When you are involved in an enterprise which makes products which do not measure up to the standards of other participants of the same industry are you doing an ethical job? When you are engaged in making a product which really serves to enrich the life of another, is this a more ethical performance than one engaged in the manufacture of such devices as illegal fireworks, Saturday night specials, or highway radar detectors? One might argue on either side as to whether individuals who are engaged in the design and manufacture of ammunition that can seriously paralyze and injure are behaving ethically. Is such an endeavor more ethical during wartime than during peacetime? There is a debate over whether the public, licensed or unlicensed, should be allowed to purchase and own assault rifles, such as the AK 47. Is this a real sporting gun?

One of the specific code references that applies to management is the responsibility for educating engineers and others in the organization. Another is the use of energy and materials in a responsible manner. How many other sections of the code for each of the engineering societies can you find which can be specifically applied to the management of an organization?

When we discuss management it may be proper to discuss management of the engineering society. An engineering society is a group of individual members. The society management should therefore be responsive to the needs of individual members. Officers, with rare exception, are supported in their roles by their employers. Therefore, it becomes necessary for officers to split their loyalties between what they may conceive as the general welfare of the society's members, the welfare of employers and their employees, and their own welfare and employment. The questions on which to vote may be difficult ones to answer, because officers see them as questions which are either more favorable to the company or engineer employees who are, in fact, the society membership. How threatening would superiors view the passing of this statement by the society management? Such questions come up when the society management is asked to take a position on such factors as the rights of employees, support of whistleblowers, and general statements of what may be an ethical or unethical activity. The same situation comes up when votes are taken in committees. A procedure often resorted to by company and government employees in such situations is either to ask that the secretary note that they had abstained from the vote or that their vote is a personal vote and does not necessarily reflect the opinion of the company or the service. Such action may or may not leave individuals open to criticism by their superiors. Fortunately such action is not too often a serious offense against the company. The competing loyalties are to the company, to the society, to the members, and to oneself.

In contrast one should examine the activities within the ASME Standards group that resulted in the Hydrolevel case described in Chapter 3. Here, the interpreters of the code were unethical in that they used their position for their own advantage and to the detriment of a competitor. This false interpretation resulted in ASME losing a large lawsuit. The charge against ASME was essentially one of failure to properly regulate and supervise. A similar charge can often be leveled against a chief executive or some lesser manager. Sometimes these failures are the result of unethical conduct, sometimes carelessness, and sometimes fraud by an underling. Almost every engineer has lost a supervisor. Can you describe the reason for the discharge of the supervisor and whether the reason was an ethical one or due to some other cause?

A group of engineers prepared a large report for an organization. The report was delivered and shortly thereafter one of the authors discovered a major error. The report was recalled and redone. Should the group have requested an increase in their fee?

A firm had a business specializing in accounting. Several of the members of the firm were licensed as CPAs. In two instances I knew their client was engaged in manufacturing a product. Each client had great difficulty and was experiencing many returns from consumers and distributors. The product did not work as it should. In fact, most of the product that reached the end of the production line did not work. To those of us in the quality business knowing the latter would make us suspect the former. We might say that there were both quality and reliability problems.

The accountants offered to help their client. They insisted that every item going down the line to final assembly and every item coming in to the plant be subjected to 100 percent inspection. The accountants reasoned that the result would be a good product at the end of the line and a good product in the hands of the customers. When I was told about these cases, at different times, I suggested that the company needed quality and/or engineering aid. "Right now they are in deep trouble, we will give you a call when things settle down," was the accountant's remark. The companies both went out of existence. The costs were astronomical and they never did get the quality problems settled. This is a situation where accountants tried to engage in quality and reliability engineering. Was it ethical of the accounting firm to even try to help these companies in this manner? Did they contribute to each company's demise?

Can you identify any of the ethical rules that were or were not observed based on the Codes of Ethics of the American Society for Quality Control and that of the internal auditors? If you have access to the Code of Ethics of the CPAs perhaps you may also use it.

5

Ethical Problems in Engineering, Private Practice, and Corporate Employment

Who is an engineer? Who is a quality control professional? The question of who can write that he or she is a "professional engineer" or a "Professional Engineer" is often a matter of state law. In some states the latter version is allowed only if one holds a state license to practice as a Professional Engineer.

In most states there are licensing procedures for lawyers, physicians, nurses, beauticians, teachers, dentists, realtors, pharmacists, and engineers. The state of California licenses engineers according to their specialty. This restricts their practice. In other states engineers are licensed as "engineers" and the restriction is self-imposed. They may practice and engage in any engineering service they feel qualified to perform.

This to some extent is true in other professions. Physicians in metropolitan areas provide services in their specialty. A physician may be an internist, an anesthetist, an allergist, an otolaryngologist, a surgeon, or have some other specialty. In a metropolitan area few internists or general practitioners would perform surgery except of the most minor type. In a rural area the physician practicing alone and with no one to consult with might find it necessary to serve as a pediatrician, a gynecologist, an obstetrician, and a surgeon.

Admittance to the specialty and the granting of the title "Fellow or Diplomate" of the specialty or its society is acquired by study and a form of apprenticeship. In a metropolitan area general practitioners would not perform elaborate surgery because they had insufficient training. The metropolitan hospital would not allow them to use its facilities for surgery. Hospitals and physicians might be the subject of malpractice suits in the event patients did not recover or developed serious complications. This could occur in the event physicians were practicing in specialties in which they were either certified or not certified.

There is also the reverse situation. Patients might sue hospitals if they were not taken care of during emergencies. This is not uncommon. In one instance a patient went to a hospital. The staff refused or neglected to take care of her promptly and properly, even though there were competent physicians available. They said the patient belonged to a group the hospital staff felt was competing with them. A malpractice suit was filed against the hospital, the group that was under contract to serve the patient, and the doctors in the hospital who might have served her. Malpractice claims, on the average, appear to run to greater sums than many product liability claims.

Some specialties in medicine, particularly orthopedic surgery and a few others have particularly high premiums for proper coverage against malpractice claims. Ethical action in medicine, properly serving patients, and maintaining expertise in a specialty is therefore no more nor less than self-protection.

In one case a woman was giving birth. The obstetrician noted that she was turning blue. The anesthetist had been derelict and allowed her to pass out. The patient was permanently brain damaged and disabled. The baby survived. There was a large settlement since three children had been deprived of their mother and the woman had to be confined to an institution. In such a case a physician may or may not lose his/her license or be unable to practice at that hospital. The information does not necessarily travel from hospital to hospital or across state lines.

The dichotomy of practice in the dental profession is less marked. Many dentists will do root canal work, extractions, and oral surgery even though there are some who practice each of these activities as a specialty.

In law there are specialists and generalists. There are some states where the practitioner is specifically licensed to practice specialties. Some firms engage in general practice, others in wills and surrogate matters and some in product liability and malpractice cases. There are also large firms that engage in corporation law.

These firms will frequently do special jobs for members of the corporations they serve including such matters as divorces and tax matters. Lawyers are admitted to the bar in a variety of ways. In many states an attorney specialist may be admitted to practice for specific cases. In most states lawyers are admitted after having completed approved law courses, passing examinations on subject matter, and also passing extensive investigations of their personal life, supposedly proving that they are ethical and moral individuals. In many states an attorney specialist from another state may be admitted to the bar to practice in a specific case.

Based on the fact that the integrity of lawyers is more seriously challenged by the examination of their life and job history and letters of recommendation from former employers, neighbors, and character witnesses, do you believe that lawyers are more ethical individuals than members of other professions?

Engineer and quality control personnel may be professionals in the sense that they are members of a technical or scientific society, they may be licensed as Professional Engineers by a state or other legally constituted licensing agency, or they may be certified as a quality or a reliability engineer by ASQC. Most medical and legal professional school graduates take and pass licensing tests, most engineers do not. According to the ABET, roughly a third of the engineering school graduates take the Engineer in Training examination. This is a national examination in the United States and makes it possible for engineers to take professional examinations some four or more years later when they have accumulated some responsible experience. If graduates do not take the Engineer in Training Examination when they finish their bachelor training or master training, they must take much more elaborate examinations when they apply for professional license. There is another accredited engineering track called engineering technologist. The percentage of these graduates that take the Engineer in Training Examination is much smaller. Approximately 75 percent of the engineering graduates taking the Engineer in Training Examination pass, and 33 percent of those graduating with degrees in engineering technology pass.

The 1986 records of resident engineers licensed as Professional Engineers in the United States is approximately 343,000. The total for all states is 596,000 but this includes registrations of people living out of their licensing state. It is possible that many engineers are not registered in the state in which they live. This would be true of the California registration of Quality Engineers and possibly some other specialties. However many engineers having registration in one state have become registered in other states where they have some practice. In general the rule is that if you maintain an office in a state you must maintain a registration in the state. Most every state has granted registration to many who obtained it in other states. In fact they have been known to grant licenses to practice as Professional Engineers to individuals who were granted California registration as Quality Engineers under grandfather clauses when the quality engineering option was first offered. The point that I am trying to establish is that most engineers do not obtain a license. There are more than 78,000 engineers and over 12,000 engineering technologists graduating each year. The total registration is thus the equivalent of approximately four years of graduating class members.

Why do so few engineers register? Perhaps it is because so many engineers work for corporations and design products or operate in an area where public safety is not thought to be an important responsibility. A professional engineering license is required for the individual who signs certain documents. When applications for building permits or major alterations to buildings or structures in which structural changes are made, a registered engineer's or architect's signature is required on the plans. Public works plans require the signature of a registered engineer certifying as to the safety of the design. Pressure vessels and reactors also require authorized signatures. When heavy equipment is installed in a structure the plans may also need approval. Ordinarily the design of consumer products does not require such a signature. In fact the design of many products can be done by people who have no concept of stress analysis. In one instance the designer and builder of a roller coaster had no engineering training whatsoever and the resulting accident seriously injured a rider.

Certification and listing schemes tend to reduce the chances that consumer products will injure people or destroy property. The Underwriters Laboratories and several others have a listing service that requires submittal of design criteria and the passing of a variety of tests for each device that is to bear the UL or other listing tag. Wires and wiring devices require this listing. There are also tests for high-voltage wires and equipment used in utility wiring. Despite such safeguards there is no 100 percent assurance that the system will not break down, as happened when the first approval was given for the use of aluminum wiring in houses and mobile homes. The connection procedure that had for long periods proved satisfactory for copper wiring did not provide a safe and reliable connection for aluminum wiring. As a result there were connections to outlets and between wires which loosened and provided a high-resistance connection which created sufficient heat to cause fires and the destruction of residences.

Another failure occurred (circa 1969) on listed television sets in which certain components created sufficient heat to ignite the set and then the surrounding areas. Some of these resulted in fatalities as well as residential damage.

The safety of most consumer products is determined by the manufacturer and not by the consumer. It is only after there have been a series of incidents involving a product that the consumer's voice may be heard, and changes or a recall initiated. There is some consistent effort on the part of listing services such as the Underwriters Laboratories to analyze the product and its design prior to its being granted a listing. Nonetheless these services generally serve manufacturers from whom they receive their fees. These agencies are honorable and must be credited with improving the safety of products. But their standards are developed by the agency and the manufacturers and not by consumer groups. There is some consumer review but this is generally done by groups of consumers who do not need to have scientific, engineering, or technical knowledge. As a result most of the press and the periodical consumer advocates are people with charismatic appeal. They do help the consumer, but this is after the fact rather than before the release of product review.

The strongest agency on the consumer's side is the Consumer Products Safety Commission (CPSC), and with all due respect, this agency's power has been limited and its activities restricted by executive department actions. The Consumer Product Safety Commission was established under Public Law 92–573 by the 92nd Congress on October 27, 1972. The CPSC, in accordance with the initial act, was to be promptly notified by manufacturers when they found that their products were unduly hazardous. This discovery was usually in the form of an untoward incident in which there was a severe injury or a serious loss. The activities of plaintiff's lawyers may at this time be a more serious threat to the well-being of any organization whose product causes injury or loss.

The job of the quality engineer in many of these instances is to examine the manufacturing and control system to find how the product escaped, in the event it was a manufacturing defect, and to close the gate on further occurrences. In the event this occurred several times it may behoove the organization to process a recall to reclaim the defective product and remove it from the market. In the event the defect was one of design and present in all product, perhaps even varying in severity, there might have to be a general recall and an attempt made to remove all dangerous product from the market by repair or replacement.

The most famous engineering failures are probably those of bridges, roofs, and other structures such as the skywalk at the Kansas City Grand Hyatt Hotel. These are due to a variety of causes, including improper design and workmanship and failure to supervise the construction which allowed for changes which violated design criteria.

The lack of engineering involvement in the stream of commerce is shown by the fact that the Institute of Electrical and Electronic Engineers has no record of any instance where the Code of Ethics was involved in a complaint to the institute.

In private practice there is another factor that may make engineers seek licensure. Billings from professionals are recognized as requiring payment and are more likely to be collectable in the event of a dispute between clients and engineers.

Private practice is more common among civil engineers than any other group. They are involved in the construction business which requires registration for plan approval. The fact that there are so many more cases of ethical review among the civil engineers may be due to the character of those involved, or due to the business situations. Perhaps there are other reasons.

The major incentives for registration as a Professional Engineer seem to be two: the need for a license to sign plans; and the desire of some engineers to enter into private practice where the holding of a license enhances ones creditability.

What work do professional engineers do in the quality field when they are engaged in private practice? We will assume that private practice is defined as setting up an office outside of a major corporation, either alone or with others, and offering services which relate to the quality field.

In such situations engineers might assist corporations in identifying the cause of a discrepancy, between the data obtained and that which they thought they should or wanted to obtain. Engineers might help organizations set up or change their manufacturing process so that the production of a nonconforming item, lot, or product was an extremely rare event, such as less than three per million. Engineers might help organizations develop experimental procedures to identify the factors which were responsible for a low process yield, or try to identify whether the cause of that low yield was internal or external.

In one instance I assisted an organization with a problem of low yield from a manufacturing line. Two pieces of glass were heated along their edges and sealed together by pressing glass. The product had previously been cylindrical in shape and had now been changed to a rough rectangular shape. With the introduction of the change the fraction of unsatisfactory seals rose markedly. There was a firm belief in the minds of the engineers and managers that the failures were due to identifiable attributes in one of the purchased items. The incoming inspection had been able to identify about a dozen characteristics and six of these were blamed for the failures. We set up some experiments and unfortunately found that after the suspected irregularities were sorted out the failure rate using only good parts looked exactly like that using all parts coming in without sorting any out. It was also discovered that running the culls through the production process yielded the same type of failure distribution as running good units through the process. There was some other factor which had not been identified. Despite the fact that this experiment did not prove the suspicion it was valuable in that it proved that the suspicion was incorrect and that some other tack needed to be taken.

A consulting engineer working with a company may do almost anything that an employed engineer might do. The big advantage that some organizations realize is the ability to have an outsider, who has more competence than the insiders, to provide some advice and guidance. There are other instances where the consulting organization provides the manpower to do peak jobs or get work out at an earlier date. The supplying of extra help over a short period is usually a function of a different type of service. The organizations who do this are sometimes called "body shops" and operate like temporary secretarial, accounting, and similar services. The personnel supplied by a body shop might or might not be registered.

Some consulting services have assisted organizations to comply with necessary requirements or to conform with standardization and registration procedures. Others specialize in training, writing specifications, and in general assisting organizations over rough passages in their governmental and industrial contracts by showing them how to properly interpret specifications and procedures. Still others have assisted in settling internal and external disputes which is sometimes more economical than going to court.

Among the tasks that consulting engineers might be requested to do is assist in finding the cause for discrepant material, or find why the process seems to work well at some moments and not as expected at others. Another task might be assisting a company in identifying and correcting the low yield of a process.

In some of these instances licensed and unlicensed engineers are involved. Licensed engineers are covered by the state licensing divisions which legally place certain additional restrictions on those that are registered.

The Statutes of the State of New Jersey, in the publication that includes the Roster of Registered Professional Engineers, lists the subject of *misconduct.* Here descriptions of prohibited acts are provided. (Figure 5.1). It should be noted that the language of the statute is specific in stating that the reporting of violations shall be made.

There are other interpretations of what is correct conduct. These opinions are not universal. [The summary of the remarks by Richard D. Wood, Chairman of the Board of Eli Lilly, to students at the Krannert School of Management at Purdue University (Figure 6.1) and the Code of Ethics of The Institute of Internal Auditors (Figure 6.3) are other views not necessarily all in agreement.] If they differ, how do they present different views?

Private practice is very common in some professions but not as common in the practice of engineering. Private practice by physicians means working alone or within small groups and taking care of patients. The physician might be an internist, an obstetrician, a psychiatrist, a surgeon, a family doctor, another specialist.

As surgeons they would take care of a large portion of their duties in hospitals or similar institutions. They must therefore have an association with the institutions.

Family practitioners may occasionally have very sick patients in a hospital. Therefore these doctors may not be associated with institutions in the same manner as the surgeons and may not have admitting privileges at the hospital. As such family practitioners cannot get sick patients into hospitals as readily. One would expect that doctors with admitting privileges or in some manner associated with hospitals would be recognized as competent by their peers. This is not always the case. There have been private hospitals where the physician owners would keep the hospital beds filled by keeping their patients there for overly long periods of time. This not only keeps the hospital operating more profitably, but also adds to the physicians' incomes since they bill for every day they visit patients. This bed filling by placing people in the hospital may be ethical as well as unethical, and in some cases the hospitals have been the very best and suited to certain kinds of treatments.

The federal government, through Medicare and Medicaid, is a major payer of hospital bills. To control the outflow of funds and also to curb the overly long stays, the government adopted a DRP, a diagnosis-related payment system. For each diagnosis of a person entering a hospital the institution receives a specific dollar amount. The payments for various classes and locations of hospitals may differ. A large urban hospital runs a larger cost than a small rural hospital. Nonetheless if a patient enters with a specific diagnosis or to receive a specific treatment, the hospital will receive the same amount whether the individual stays for a period of time shorter or longer than the estimated average. This encourages hospitals to make patient stays shorter and discharge sooner. This is not all good or bad. Hospitals are not healthy places to be. It has been found that there are many infections which pass from person to person in a hospital, and therefore the sooner a person can be discharged the better his or her chances of not catching an infections. Some 40 years ago the average stay for an uncomplicated maternity case was two weeks. today many maternity cases see the mother and baby leaving the hospital in one or two days. This is considered good medicine and it not only may be healthier for the mother and the baby, but it is also beneficial on a cost basis for the government and the health insurance industry. It is also beneficial to the subscriber because it reduces premiums. The physician can also be placed in the position of not being able to collect excess fees. However in maternity cases many physicians have flat fees for prenatal and birthing services. Obstetrics calls for the presence of the physician at all times of the day and night. The birthing event does not respect time and thus it is not unusual for several specialists to pool their interests so that one can take a weekend off or a vacation while others are available to take care of a birth when needed. Such group practice requires that each of the group be familiar with the other's patients and practice and be aware of the unusual. There are other types of group practice where one or more are internists or family practitioners and others are specialists, who can take care of the services that the patients may need as well as handling specialty services for other physicians. All of these doctors may be said to be in private practice. They are individual or group entrepreneurs. Other physicians work for drug companies, and are physicians for company employees but they represent the company first. Still others work for health maintenance organizations (HMOs) which are a form of insurance company and so classed by many states. These organizations accept a monthly fee for service and supply all service including office visits and hospitalizations. There are a great many variations in the degree and completeness of service just as there are many variations in insurance plans. The subscriber visits the site, and is served by the physicians employed there. There are also prepaid service arrangements furnished by physicians. The HMO and other groups know it is to their advantage to provide the shortest hospital stay and minimize the number of visits. This is at variance with the thinking of some fellow physicians who maximize visits to maximize fees. There are some excellent services available from all of these sources, and as every quality professional knows, there are good, average, and poor of almost every category. The same is true of the medical advice and treatment one can receive from any type of service.

Since the consumer revolution of the late 1960s, physicians have been subjected to malpractice suits. A large number of these cases have been settled both in and out of court. Some of these settlements have been large. This makes the malpractice insurance premiums some physicians pay extremely large. Physicians who make mistakes and treat their patients improperly may be successfully sued by patients or their survivors. This causes doctors to practice defensive medicine and perhaps require more than the minimum number of tests to assure themselves that they do not make errors. This runs up the cost of treatment.

Another way of avoiding errors and successful malpractice suits is to obtain the opinion of another physician or a specialist. Some insurance companies pay for such consultation.

The physicians who work for groups such as HMOs may or may not be thought of as being in private practice. One criterion of private practice might be the matter of whether practitioners take their clients or patients with them when they move from one location to another. Physicians working for HMOs would not be likely to do this. There are similarities with other professions.

The code of ethics for physicians is published by the American Medical Association. It specifically points out that the physician has the right to refuse to serve patients except in the case of an emergency. It is like the engineer's code in that it requires the physician to serve society. (The AMA code is included in Chapter 2.)

Engineers in private practice might band together to form a corporation, a partnership, or another organization. The restrictions on these organizations is not consistent in all states. There are different rules that apply when the corporation is attempting to practice professional engineering and when the consolidation is for other purposes. There are large engineering organizations doing design and construction and supervising large projects. The engineers in these organizations are in some instances owners, in others, partners and bosses, and in some instances employees. There may be dual responsibilities—to the firm and to the client.

Again these engineering organizations might take on a large project for government agencies or corporations. These programs might include training the employees of an organization in quality methodology, assisting them in solving the problems of the corporation, or even solving the quality problems on the outside and then introducing corrective action into the organization.

Another task that a consulting organization might undertake is to write the procedures for agencies or corporations. The reduction to written form in one compendium of the many rules and regulations covering the construction or operation of a power plant, or a nuclear or fossil fuel generating system design might be contracted out, particularly if it is felt that this is a major job and there is no need to expand the staff.

Still another task might be to set up a program of audit and certification by a group representing major companies in an industry. This might include the adoption of the laws and regulations that govern a process. The consulting organization might then furnish the audit personnel to review and certify the plant operations and contract with laboratories to perform tests on samples picked up during the audits. In this way a certification scheme might be developed which was sufficiently satisfactory to induce the federal or state agencies to reduce the frequency of their inspections or to forego inspections. Other organizations might just supply personnel as was previously mentioned for systems like temporary secretarial and accounting firms.

Lawyers have similar groups and arrangements. Some lawyers practice as individuals, some for law firms which may run as large as several hundred lawyers, and some work for corporations as the legal counsel but in reality are members of the staff or officers of the corporation. These lawyers are not in the private practice of law. Another criterion of private practice might be the question of whether the individual has a boss and whether he or she shares in the profits of the organization or is on a salary.

The legal profession is also very much older than the engineering profession in the sense of organization. While we saw a reference to a code of ethics for physicians going back to the time of Hippocrates for the medical profession (see Chapter 2), the legal profession in the United States has a code of ethics and ethical procedures which are more than 100 years old in their antecedents.

According to the bulletins of the American Bar Association the legal Canons of Ethics are traceable to the lectures of Judge George Sherwood (1854), the subsequent writings of David Hoffman, *A Course of Legal Study* (2nd edition 1836), and the code of ethics adopted by the Alabama State Bar Association in 1887. The canons as of 1985 were published as the *Model Code of Professional Responsibility and Code of Judicial Conduct*. The code has been extensively revised and edited so that the eight major canons are short, but each is followed with ethical considerations, disciplinary rules, and notes indicating where the specific citations come from, plus an index which takes up more than 100 pages. These same canons are then adopted by individual state bar associations, and when the state statutes which support the various opinions are included, the book gets to be a very extensive one. This is particularly interesting because each canon is discussed in depth and the various connotations have been explored by various authorities in different instances. The engineering codes of ethics do not seem to have accumulated a similar background.

The previous publication is still the basis for many state codes. The more recent code revisions have been adopted by approximately half of the state bar associations. It is entitled *Model Rules of Professional Conduct and Code of Judicial Conduct* (1983). The older code *Model Code of Professional Responsibility and Code of Judicial Conduct* (1970) is easier to quote from since it is more terse. Some of the canons are quite similar to those that the engineering societies invoke. One that is not in the group espoused by engineering societies relates to making services available to the indigent.

The U.S. Supreme Court's Miranda ruling is related to the conviction of criminals without what was determined to be due process in their defense. After the ruling there were many appeals that were pending by those who believed themselves to have been improperly convicted. the backlog was great and many of those making the appeals were indigent. The bar associations asked many law firms, some of whom were not engaged in the practice of criminal law, to accept one case and process it. I knew of one patent counsel who won the appeal for the client who had been assigned to them.

Among the problems that are common to the legal and engineering professions are conflict of interests. The legal canons explore such items as the representation of an individual who is pressing litigation against a party who was a former client of the attorney or whom the attorney had worked for on a similar cause. The canons also discuss working for a government agency and a private party, successively, presumably when the subject matter is related.

Engineers in private practice have loyalty and ethics problems when they attempt to serve two or more clients in the same industry. If such an event were to occur it would behoove the engineer as well as the lawyer to disclose to clients the employment by other companies. Sometimes there is a mutual interest, and sometimes there is a situation in which the companies are competing. In the first case there is justification for an engineer working for several companies, whereas in the latter case it would be difficult.

When operating in private practice engineers are often asked to sign nondisclosure agreements stating that they will not disclose company business to another firm. In many instances this is quite necessary since the activity represents some product or service the company is about to market. Of course if engineers were to disclose someone else's secrets, the recipient would doubt the integrity of the engineers and not trust them. The disclosure would be tantamount to self destruction.

Quality engineers or practitioners have another problem. Having taught one manufacturer how to produce a widget so well that there were only two or three nonconforming units produced per million, are they free to teach someone else in the same line of business to do the same? Is this a situation where they can never work for another company in the same line of work or is there a time limit? Would it be proper to work for the competitor a year later or two years or five years or 10 years later?

Bridge designing civil engineers could be in very different situations. They learn more about bridge design as they progress. The next bridge is not in competition with the earlier ones, and therefore there is no time limit that ought to exist between one job and the next. When an employed engineer leaves one company and goes to another making product that competes with the prior employer's, there are problems. How long must he or she wait to disclose strategies learned at the last employer's plant? This type of problem is often covered by employment agreements which prohibit an engineer from accepting employment with a competitor for a period of time. Such covenants, when imposed forever, have often been declared nonbinding. When they are for periods of time as short as a year they have often been enforced.

The agreements as to confidentiality are also most important for the short term. When you have worked for Company A and then work for Company B, with a long period between employments, the likelihood of divulging material that could be detrimental is remote.

What are the canons of the engineering codes of ethics that an engineer in private practice ought to examine before agreeing to work on a subject that is closely related to or the same as that which a former or present client has or had an interest? Perhaps the engineer ought to pay particular attention to public safety and welfare, loyalty to more than one client, and the task of serving society and keeping himself or herself educated. To what other portions of the code of ethics ought he or she pay particular attention?

One question that can arise is what should engineers do when they have, in their opinion, properly advised a client and then discover that the client has not taken the advised action? In some instances the question is moot. The client discovers that the action taken is wrong and adopts the action that was suggested. Let us say that a client employs an engineer to help determine the cause of an accident which had the potential to cause injury, and which if common to all the units because of the design, could cause injury in future instances. Let it be further assumed that this product is covered by a federal ordinance. The organization is required to report the incident and its cause to a federal agency within a very short period after discovering the cause. Then they must notify the public and recall, replace, or fix the units in the field in a manner that will markedly reduce or eliminate the hazard. If the cause can be identified as likely to have been present in a small portion of the production, then those units may be recalled instead of the entire production. The engineer feels that there is a hazard. The company is not convinced, but is certain that there is no hazard, or at least they so state. The company will not report this. Should the engineer? Should he or she discuss this with the organization? Should he or she cease working with the organization?

There are similar situations. When you are in a plant you may note conditions which are unrelated to an assignment. While working in an automobile plant in

Egypt, I was assigned the task of making recommendations on what should be done to improve the quality of the process to reduce internal costs and produce better-quality vehicles. During this task I noted a hazardous situation. During the oral delivery to the board I stated that, though it was outside of my assignment, it would be wrong not to call this hazardous condition to the attention of management. The chairman commented that he had seen this but thanked me for the comment and promised that the situation would be corrected, he hoped, before someone was injured. I heard at a later time that nothing was done.

One must wonder whether codes of ethics are ever enforced. It must also be evident from a statistical point of view that in any profession there are those that are outstandingly good, above average, average, below average, and incompetent. The reason for their incompetence may be any number of factors including dishonesty or an unwillingness to work effectively, illness, alcohol, or drug use, or sometimes just incompetence. This can be a serious problem for the individual who employs or engages the incompetent. The board that licenses these people in some instances is expected to censure the individuals who do not practice ethically. The failure to practice ethically may take several forms. Dishonesty, fraud, and failure to perform in a manner consistent with the public welfare are all just causes. Revocation of license is relatively rare.

Richard Greene writing in *Forbes* (October 5, 1987) reported that the medical profession is estimated, by the *New England Journal of Medicine*, to have 20,000 physicians, approximately 5 percent of the total, who ought not to be practicing [15]. In 1985 only 2,108 physicians or 0.4 percent were subject to serious disciplinary action and only 406 had their licenses revoked. This number is up from previous years. Greene goes on to say that one of the reasons so few actions are taken by state boards is that the board members are afraid of being sued even when they are correct in their judgment. Several have had to spend large personal sums to defend lawsuits initiated by individuals whose license to practice had been revoked.

The chances of disbarment of a lawyer may be even less, and the suspension of or removal of a society member is probably even smaller. The number of cases where members' credentials were reviewed and their suspension from or removal from a society was discussed, is very small. The chance of suspension is small except possibly for the most heinous offense.

I have asked several societies for statistics on the number of members they have suspended, disciplined, or removed from their roster due to flagrant abuse of their ethical codes. The societies report no activity in their censure or suspension of members with the exception of the American Society of Civil Engineers, who have a continuing effort with quite a few cases of ethical violations each year. (See Chapter 12.) The American Society of Mechanical Engineers had supplied no information as of this writing.

Another facet of the professional engineer's responsibility is how best to work when the task involves the work of other engineers as well as his or her work, and how to care for the welfare and well-being of the client.

Figure 5.1 STATE OF NEW JERSEY STATUTES, ADMINISTRATIVE RULES AND REGULATIONS

Subchapter 3. Misconduct

13:40–3.1 Enumeration of Prohibited Acts

(a) Misconduct in the practice of professional engineering or land surveying shall include, without limitation:

1. Acting for his client or employer in professional matters otherwise than as a faithful agent or trustee; accepting any remuneration other than his stated recompense for services rendered.
2. Disregarding the safety, health and welfare of the public in the performance of his professional duties; preparing or signing and sealing plans, surveys or specifications which are not of a safe design and in conformity with accepted standards. If the client or employer insists on such conducts, the licensee shall notify the proper authorities and withdraw from further service on the project.
3. Advertising his work or merit using claims of superiority which cannot be substantiated.
4. Engaging in any activity which involves him in a conflict of interest, including without limitation:
   i. A licensee shall inform his client or employer of any busi ness connection, interest or circumstance which might be deemed as influencing his judgment or the quality of his services to the client or employer.
   ii. When in public service as a member, advisor, or employee of a governmental agency, a licensee shall not participate in the deliberations or actions of such agency with respect to services rendered or to be rendered by the licensee or any firm or organization with which he is associated in private practice.
   iii. A licensee shall not solicit or accept a professional contract from a governmental agency upon which a principal, officer or employee of his firm or organization serves as a member, advisor or employee.
   iv. A licensee shall not accept compensation or remuneration, financial or otherwise, from more than one interested party

for the same service or for services pertaining to the same work, unless there has been full disclosure to and consent by all interested parties; except, however, that notwithstanding such disclosure and consent, a licensee shall not approve his own plans or work.

v. A licensee shall not accept compensation or remuneration, financial or otherwise, from material or equipment suppliers for specifying their product.

vi. A licensee shall not accept commissions or allowances, directly or indirectly, from contractors or other persons dealing with his client or employer in connection with work for which he is responsible to the client or employer.

5. Affixing his seal to any plans, specifications, plats or reports or surveys which were not prepared by him or under his supervision by his employees or subordinates.
6. Violate the engineering or land surveying laws of any state.
7. Permitting or allowing any person not appropriately licensed pursuant to N.J.S.A. 45:8–27 to act for or on behalf of the licensee as his representative, surrogate or agent while appearing before any public or private body for the purpose of rendering professional engineering or land surveyor services.
8. Failure to determine and document the identity of the client prior to commencing any work. All correspondence, contracts, bills shall be addressed to that client, unless expressly directed otherwise, in writing, by the client.*
9. Failure to keep a client reasonably informed about the status of a matter and promptly comply with reasonable requests for information.*

*As Amended, 19 N. J. R. 851(a), effective September 8, 1987

13:40–3.2 Reporting Incidents of Professional Misconduct

If a licensee has knowledge or reason to believe that another person or firm may be in violation of or has violated any of the statutes or rules administered by the Board of Professional Engineers and Land Surveyors, he or she shall present such information to the Board in writing and shall cooperate with the Board in furnishing such information or assistance as may be required by the Board.

(From *Revised Statutes—Administrative Rules and Regulations* [Trenton, New Jersey, New Jersey State Board of Professsional Engineers and Land Surveyors], Nov. 30, 1986.)

Figure 5.2: AMERICAN BAR ASSOCIATION

MODEL CODE OF PROFESSIONAL RESPONSIBILITY

CANON 1. A lawyer should assist in maintaining the integrity and competence of the legal profession.

CANON 2. A lawyer should assist the legal profession in fulfilling its duty to make legal counsel available.

CANON 3. A lawyer should assist in preventing the unauthorized practice of law.

CANON 4. A lawyer should preserve the confidences and secrets of a client.

CANON 5. A lawyer should exercise independent professional judgment on behalf of a client.

CANON 6. A lawyer should represent a client competently.

CANON 7. A lawyer should represent a client zealously within the bounds of the law.

CANON 8. A lawyer should assist in improving the legal system.

CANON 9. A lawyer should avoid even the appearance of professional impropriety.

6

History of Ethics and Problems

Ethics can be considered as part of the proper study of human conduct. Ethics and moral principles should really not be in conflict. Philosophers have written about ethical and moral values and conduct for as long as we can remember. The early works go back to the biblical writers and others, e.g., Plato. In each culture there has been a host of writers who have studied these subjects with the concept that what is morally correct is what is most beneficial to the greatest number of people. Ethical behavior being defined as the rules pertaining to a group should then be the set of rules which is most beneficial to that group and to improving its image in the eyes of the general population. Writing and adopting a set of ethical practices, each professional group, including physicians, editors, engineers, lawyers, teachers, accountants, auditors, quality personnel, and others, has in general adopted a code of ethics in an attempt to appear professional and to provide guidelines for its members. Morality, on the other hand, is often considered to be the rules that some outer and greater force set down under which people shall guide their conduct.

Ethical standards are intended to be utilitarian and provide guidance for the members of the profession or group. The rules are intended to provide the greatest good for the greatest number of members of the professions; there are other codes of ethics that are dictated by management in large companies. Where it has been felt that the conduct of one or more employee groups must be systematized and set forth in a code. These standards also direct that certain material available to the group will not be misused and will not be handled in a manner other than the way management believes it should.

Perhaps some of the ethical rules established by corporate management to protect the company may differ from the ethical codes set up by engineering and other societal groups. The documents that are included here are all quite different. One is an address by the president of a company to a graduating class of a university and another is a set of rules set forth for a group of internal auditors within a company. These rules may also be different from the set adopted by the Society of Internal Auditors as their code of ethics. The three items should be compared with the promulgations of the codes of ethics adopted by the engineering societies as given elsewhere in this volume. If there are differences between the documents and their intent, you might ask why (see Figs. 6.1, 6.2, 6.3).

As noted earlier, ethical codes are intended to be utilitarian and provide the greatest good for the greatest number. The effort here is not to try to define what is good and what is bad, but to identify efforts on the part of various members of the engineering/industrial complex to define ethical behavior by means of an action, such as developing a code of ethics and then enforcing it.

Confining ethical standards and behavior to professional groups may be too restrictive. Business and industrial groups also should be ethical. Most of these groups are required to follow the laws of the land in which they operate. The violations of a code of ethics which are likely to be serious and to be acted upon involve human safety, welfare, and money.

When physicians do not provide their clients with the most appropriate medical care and patients suffer loss of function or loss of life or do not recover fully or as rapidly as they probably would have, had they received proper care, there has been a violation of the code of ethics. This does not necessarily indicate any interest on the part of physicians in obtaining larger fees or in making more visits at a set fee. It may be the result of laziness, ignorance, carelessness, or being under the influence of a drug. Nonetheless, the action of giving less than standard service is unforgivable. It must be remembered that what is the best treatment for most is not necessarily the best treatment for each individual. Some patients need different care and some have adverse reactions to certain treatments, medication, and materials. However, we expect out physicians to be aware of our problems and to treat us properly. It might be unethical of patients not to regularly visit physicians so that they are both aware of any peculiarities that exist.

On the other hand, physicians keeping patients in the hospital longer than necessary is not only denying beds to other patients who may be in greater need of service, but is also incurring larger hospital bills for patients or their sponsors. Physicians may also be increasing their bills which include visiting of patients daily.

Similarly, when any professional chooses a treatment solely on the basis of its providing the most income to the professional, that may also be considered unethical.

In engineering, medicine, and other professions there is a constant change in the state of the art. Maladies that were not treatable a few years ago are now readily treatable, and the recovery rate is extremely high. The fatality rate for many diseases has dropped. People live longer. What might have been the accepted treatment 50 years ago or even 25 years back may no longer be the best modern methods. On the other hand it may still be the best.

Structural, mechanical, electrical, and other designs have also changed. Appliances of today, even though they may resemble appliances of 20 years ago, are in general superior. They may be more efficient and they may be more complicated. In some ways they perform better and in some ways, because of the use of sophisticated electronics, it is more difficult to analyze the cause of breakdowns and to make simple, low-cost repairs.

For example, the methods of calculating used today are so much more powerful than those available 25 years ago, that there is little excuse for not doing calculations by the proper method rather than using approximations. Hand-held calculators and small computers can do calculations in moments, that 20 years back, would have taken most engineers days, weeks, or months to do. In the olden days there was great demand for methods of approximations.

Then, as now, the conservative engineer uses a factor of safety when designing a bridge, structure, or device. This factor, sometimes referred to as a factor of ignorance due to the fact that exact strength of the material, the exact load, and the exact use, are not known, must be larger when the calculations are crude than when the calculations are precise. In olden times the practice was to greatly overdesign. As a result, the bridge, machine, or product is still in service. This is true of some buildings and bridges that were built thousands of years ago. Today it is impossible to build with such large factors of safety because of economic pressure. The customer is not interested in investing in something that will last forever. Bridges, even when they are in good condition, become too small to carry modern traffic. Due to present road design procedures, these bridges may also be in the wrong location and sometimes are unable to carry the weight of vehicles. What was once a solid house with stone walls and oversize timbers to support the floor is thermally unsuited to today's fuel costs. The requirements of a changing technology and the cost structure cause a great variety of products and buildings to become obsolete. New models and designs are more useful and displace the old even though the older unit may perform as well as it did when it was new.

Present-day telephone designers have more than halved the life requirements used by their predecessors, since they feel that the units will be replaced long before they wear out because of the advantages of newer designs. A few decades back designers expected such equipment to last until it wore out and therefore insisted on designs which would have a longer life.

For years, Europe has rehabilitated its old buildings, while here in the United States we have a habit of tearing down and replacing. Canada reinsulated many of its buildings to reduce the use of fossil fuel. For several years the U.S. government provided tax incentives for installing more effective thermal insulation to reduce fuel usage.

Waste may be unethical. Using fuel wastefully creates additional acid rain and increases the greenhouse effect. Is it unethical to contribute to waste when there are alternatives? One is nuclear power. Europe and Russia are big users of nuclear generation systems. The United States has only a small portion of its power needs supplied by nuclear energy. Is this unethical or are there other problems? What sort of balance is proper between the risks of using and not using nuclear generation as the prime method of providing electric power?

There is a long history of catastrophes that can be attributed to the failure of man-made structures and equipments. Some of the following catastrophes were precipitated by a natural phenomenon:

Wasa sinking - 1628
San Francisco earthquake - 1906
Tokyo earthquake - 1923
Tacoma Narrows Bridge collapse - 1940
Columbia Exposition fires - 1893
Mississippi River ship explosions - 1838–1850
Titanic sinking - 1912
Kansas City Skywalk collapse - 1980
Hartford Civic Center roof collapse - 1978
Submarine *Thresher* loss - 1963
Chernobyl nuclear plant failure - 1986
Challenger loss - 1986
Tower of Pisa - 1200
Black Death and Bubonic Plagues - 400; 1665; 1900
Poliomyelitis epidemics - 1915–1955
L'Ambiance Plaza collapse - 1987

The fact that many of these occurred in the United States may be due to the better news coverage that is given local events. A disaster in a remote corner of the world may not get the publicity of a local event and rapidly fades from view.

The *Wasa* was a 17th century disaster. In the early part of the century Sweden was the greatest of naval powers. The Swedish navy built a new flagship for the fleet. The *Wasa* was commissioned, launched, and set sail in 1628 to lead the fleet out of the Stockholm harbor. It was hardly under sail when a light gust came up and overturned the ship. It sank to the bottom of the harbor, with much loss of life.

More than 300 years later in 1961, the *Wasa* was raised along with more than 20,000 certified artifacts that had become detached from the hull. Since the Stockholm harbor is a mixture of salt and fresh water, the wood was not as badly deteriorated as it would have been in a more hostile environment. The material was transported to the shore in a special dry dock and surrounded by a temperature and humidity controlled enclosure. The wood was treated with a polypropylene glycol to replace the moisture in it and to prevent it from disintegrating.

Today one can visit the *Wasa* and view what a ship of that time really looked like. The gigantic jigsaw task of assembling many of the pieces has resulted in a recognizable hull with many of the fancy carvings and ornaments that were in use at that time. It was really a remarkable operation.

The reason for mentioning this disaster is the history of the *Wasa* at the time it was built and sank. The admiralty was much upset by the incident and initiated an inquiry into the actions of the naval architect. He was to be given a fair trial before any action was taken. During the inquiry, the admiral of the fleet testified that a stability test had been performed when the hull was launched. The test consisted of having about 30 men run from side to side of the deck to unbalance the hull. The test was terminated after the third crossing for fear that the hull would capsize. Normally the operation can be continued for many crossings without causing a hull to overturn. Despite this indication of instability, no corrective action was taken. The outfitting of the ship was aggressively pursued. There was no further action taken by the court of inquiry.

Little has changed in the last 350 years. Some organizations are acting the way the Swedish navy did in the 1600s by releasing product after similar unsuccessful trials.

What may be common to many of these disasters is the inability to foresee the result of shoddy engineering or manufacture, or the failure to develop, adopt, or enforce codes.

In many cases the uninitiated buy structures or devices for lower prices on the basis that more expensive is not necessarily better. They have no way of knowing nor do they ask whether alternatives or more costly designs are safer, and if so why.

Ethical, and particularly unethical, behavior is so common to the news media these days that we are almost unaware of the fact that there is a long history of writings and actions on the part of philosophers, scientists, physicians, and engineers. An individual once said that before you take any action you ought to think about it and ask yourself whether you will be proud to go home and tell your family what you did and why you did it. Will you proudly tell your grandchildren of the actions you took, why you took them, and what happened as a result?

The *New York Times Magazine* of July 24, 1988 has two articles in it which are relevant to this subject. One, entitled "Into the Mouths of Babes" [16] tells the story of Beech-Nut selling apple juice for babies which in fact was not apple juice but colored and sweetened water. Beech-Nut was owned by Nestlé. The president and vice-president of the Beech-Nut Nutrition Corporation were each sentenced to fines and a year in jail. The company had pleaded guilty to 215 counts of violating the federal food and drug laws and had been fined $2 million. Jerome J. LiCari, head of Beech-Nut's research and development had become suspicious of the adulteration and reportedly was threatened that he would be fired unless he kept quiet about the matter and did not complain. When it became obvious that the matter would come to a head and that the company might get into trouble and that the stock might be seized by the government because of the violation of the law, the management team shipped as much as possible overseas since the federal law is not violated by overseas shipment.

Those who are interested can refer to this article and decide for themselves as to the ethics of the actions of the Nestlé Corporation, Niels L. Hoyvald, Beech-Nut's ex-president, John F. Lavery, Beech-Nut's ex-vice president, and LiCari. The *Times* article was written by James Traub, a writer on legal issues.

What each of us should do is to imagine ourselves in the place of each of the individuals and organizations and ask what it would have cost us to take a different action and then ask would we have taken that action. For example as the Nestlé manager, would we have encouraged, countenanced, or allowed the shipment of the material which was in question or would we have had it destroyed? Would we have defended the case in court and the actions of the president and vice president of Beech-Nut?

Would we as president or vice president have acted at an earlier date or would we have questioned the material when the first suspicions arose? This is an interesting case not only from the legal standpoint but also from the monetary. The sweet water cost less and was not harmful. It was a case of mislabeling. Would an action like this tend in the long run to make you question whether you should buy this brand or some other brand for your infant? This seems to be a classical case where turning a substantial profit interfered with the ethical performance on the job. Is it often that money and ethics point in different directions?

The individuals responsible for this action of mislabeling and falsification of product were convicted, but the internal discovery of the activity preceded the government action. Suppose that you had been an officer of the company and it was reported to you that your company was actually selling sugar water and labeling it apple juice, what would have been your first action? What would you have considered to be the appropriate action? Suppose that in addition to receiving the report, you also found that there were millions of these jars of sugar water in warehouses and on grocers shelves. Would this have tempered your decision? As the officer of a parent company, what would your reaction have been? Under what conditions would you have recommended that the company pay all litigation costs in defending the individuals that were responsible for the generation and sale of the adulterated product? Would you continue to employ these individuals in your company?

The June 1989 *Consumer Reports* contained a follow-up on the Beech-Nut apple juice case [17]. It reported that on March 29 a federal appeals court dismissed all 359 convictions of violations of the Federal Food, Drug and Cosmetic Act against Neils L. Hoyvald on a technicality. The court ruled that the case had been brought in the wrong court; it did not rule on the substance of the charges.

The court also dismissed 429 counts against John F. Lavery the former vice president for operations. The court affirmed his convictions for mail fraud and conspiracy. Hoyvald still faces similar charges. The Department of Justice may appeal the decision.

The other article in the *New York Times Magazine* of July 24, 1988 is by William Ira Bennett, editor of the *Harvard Medical School Health Letter* [18]. The title is the "Pluses of Malpractice Suits." The author says "an important aspect of the threat of litigation is that it can prompt a useful examination of the practice of medicine itself."

Failures, and the threat of failures, can be useful in helping to improve and assure the safety of a project. It is a common practice in engineering to overdesign or provide greater strength than needed to bear the load which a structure is expected to support. A factor of safety (designing with excess strength) takes care of the chance that the material or the workmanship is not exactly the way it ought to be. If a bridge is designed to hold a string of trucks of a specific weight there is no assurance that some years later someone will not run a longer string of heavier trucks across the span. These unknown factors are taken care of by overdesign. Economically, there is a limit. It is obviously impossible to guard against the loading of 100 times the weight that is used for design.

The Hood Canal Bridge in Washington State rests on concrete floats. It was predicted that there would be conditions about once in 100 years that would destroy the span. This happened. Fortunately there was no loss of life. However there were inconveniences. The same construction features were used in replacing the bridge. Was this an ethical procedure? (For more details see Chapter 7.)

When the history of electrical power distribution is examined, it is found that the practices surrounding the early distribution systems were rapidly found to be unsafe. It has been said that the World Columbian Exposition held in Chicago in 1893 gave the world its first real look at the possibility of electrical lighting. The public was dazzled by the use of electric lighting, but there was even more amazement and surprise at the frequency of and the brilliance of the many fires started by the electrical distribution system set up on the site.

The first power-generating system had preceded the Columbian Exposition. Thomas A. Edison set up the Pearl Street generating plant in 1881. This generated direct current which was distributed in the lower part of Manhattan Island, the downtown part of New York City. This system generated 120 volts plus and minus either side of a neutral, providing a system for the use of lights and household outlets at 120 volts and providing 240 volts for larger motors. This type of distribution system was still in use in part of New York City into the 1940s and possibly into the 1950s. The danger associated with the fires that earlier electrical systems started was evidence that there was need for a codification of the practices associated with the distribution of electricity. The effort led in two directions. One was the development of the practices necessary for the safety of the individual and property which was initiated by the Fire Engineers. The other was the establishment of a laboratory to test and approve products which were to be used in the installation. These approvals were for wire, connectors, connection boxes, outlets, sockets, switches, and other equipment.

This was the beginning of the Underwriters Laboratories in 1894. The first code of electrical installation was issued in 1895. It had been in progress since the National Association of Fire Engineers met in Richmond, Virginia in 1881. The National Electrical Code (NEC) is currently published as American National Standard ANSI/NFPA 70 and is under the jurisdiction of the National Fire Protection Association.

In 1911 the National Electrical Code came under the control of the NFPA, the National Fire Prevention Association, and in 1920 it became established as a National Standard under the American Standards Association (ASA), an early name of what is now the American National Standards Institute. The work on this code is assigned to the C1 committee.

Today the National Electrical Code is adopted into law in many localities. In others it is only partially copied and altered to provide city codes such as exist in New York and Los Angeles. The code is the model for the minimum basic safe installation of electrical systems in buildings. There are sections of the code that apply to one-and two-family buildings, and other sections that apply to more complicated installations.

What does this discussion have to do with ethics? It is ethical to set up a practice that protects the home and the individual. The code not only presents minimum safeguards to protect from fire, but also from shock, electrocution, and electrical burns. It might be considered unethical not to follow the dictates of the electrical code in areas which do not enforce the code. It would certainly be illegal to install an electrical system that did not comply with the code in an area where the code had been adopted into law. It might be unethical to install an electrical system that met the minimum requirements of the code in an area where judgment should have told the engineer that the conditions were hazardous and that an analysis should be made of the necessary requirements for building a system that exceeded the minimum requirements of the code. The code treats hazardous locations and imposes special requirements for safety.

I have known sites where an engineer was called in to survey an electrical system and upon viewing it required that the system be disconnected and the building shutdown because of the imminent danger of fire.

Ethically should this have been allowed to exist? Who would have the responsibility of making the decision to have the survey made to assure safety?

Insurance companies and building departments in many localities require that a building have a certificate of occupancy before anyone can move in and live there. If properly done, this is a reasonable safety precaution. It is also a reasonable health precaution since plumbing codes are designed for health just as electrical and building codes are designed to provide for safe buildings. Considering the fact that we have had so many major failures, the Kansas City Grand Hyatt Skywalk, the various roof failures, and the collapse of buildings during construction, it is apparent that there just are not enough checks made on some structures during design and construction. The major failures and incidents which make the headlines represent a small fraction of the incidents in which people are injured or killed. The first steel skeleton structure, the Eiffel Tower, is still standing 100 years after erection despite the fact that it was built as a temporary structure. The foundation is sinking slightly due to the tremendous weight of the structure. There are many places where the tower is rusting out. The rusted portions have been replaced. In addition the structure has been lightened by removal of many stiffeners and sections which were grossly overdesigned in the 1800s because there was insufficient knowledge of the theory of structures. The errors were made on the side of providing excess strength rather than just meeting the necessary requirements.

Ethical considerations require that engineers have several responses to codes. First, follow the code; secondly, when in doubt, request an interpretation of the code; and thirdly suggest changes and improvements in the code when it is clear that the code can be misinterpreted to allow for an unsafe installation. It would be interesting to review codes to uncover other ethical questions.

Figure 6.1: The Following Article is a Summary of Remarks by Richard D. Wood, Board Chairman, Eli Lilly & Co., to Students at the Krannert Schools of Management at Purdue University

Ethics in business—just as in religion, education, and government—are critically important to the achievement of success over the long term.

Business, particularly international business, is completely interwoven with society, which dictates the rules by which business conducts itself. Society ultimately rewards those companies that play by the rules of the game and penalizes those that do not. The same rules apply to individuals as well as to organizations.

Origins of High Standards

Throughout my career at Lilly, I have observed how ethical standards can be applied and how they can produce great benefits to a business over a long period of time. Lilly was operated much as a family business until the 1960s and early 1970s when it emerged as a large, publicly owned enterprise. The Lilly family set the company's attitudes and philosophies in place at the very beginning. Their standards were so high that they were not only most unusual for their time but they could only be called visionary. Much of the company's success has to be attributed to the establishment of those attitudes and philosophies very early in the company's history—and the fact that the organization has adhered to them rigorously over the years.

Fundamental to the company's operations have been its policies involving employees. The company directed that the dignity of the individual was to be emphasized, stressed the importance of having a high personal value system based on common sense, and, above all, demanded absolute integrity in all areas of the business.

The company stated *in print*, at a very early date, that "no Lilly employees should do anything or be expected to take any action that they would be ashamed to explain to their family or friends." In today's world, that perhaps is not a sophisticated way to state a policy, but it still says a lot.

Impact on Corporate Activities

Very fundamentally, it was also directed that the company concentrate on scientific research, so that it would produce products whose claims were based on fact rather than an advertiser's creativity. Research efforts were to be as exhaustive as possible, in keeping with the current state of the scientific arts.

To their production people, the Lilly family left no doubt that their mission was to manufacture products that had only the highest possible standards. Consistency and reliability of products were absolutely essential. If a production lot didn't pass exhaustive testing, it was either reworked or destroyed.

To their marketing people, the Lillys stressed the importance of presenting balanced, honest, straight-forward promotional efforts that did not overstate the value of products. Listing the disadvantages of a product, along with its benefits, is a difficult strategy for sales people to swallow, but that was precisely the policy.

Start Early

How corporations achieve success and build good reputations over their lifetimes is a fascinating study in and of itself. But it seems to me that, in the development of a business, matters of ethics or codes of conduct ought to be put into place very early on.

And I would conclude that the most successful companies appear to be those that have had a clear focus regarding their mission and the path they wish to follow to realize that mission. Their standards of operation generally have been at a very high level, and the businesses have adhered to those standards, regardless of the consequences.

Adhering to a set of principles if difficult. Human events move swiftly. Products change. Competitors change. Markets change. Laws, regulation, and people's attitudes are modified.

Reflection of Personal Values

While it makes good sense to keep an eye on a changing environment, one must resist making changes in a set of principles just for the sake of change. Colonel Eli Lilly once said, "Do nothing that you would be ashamed to explain to your family or friends." Ethical standards in business are nothing more than a reflection of the personal values of the people in an organization.

Now, let me hasten to say that no corporation has a perfect track record in this respect. Ours does not. I suspect that any company goes off the rails occasionally—usually because of aberrations in individual human behavior rather than a defect in broad corporate intent.

As students begin to think of their careers, I would urge them to seek an organization of which they can be proud, not only in terms of its financial accomplishments in the marketplace, but most importantly, in *how* it goes about its business.

(Reprinted from: Eli Lilly and Company First Quarter Report, 1988, with the permission of Eli Lilly & Company.)

Figure 6.2: Rockwell International, Inc.

PROCUREMENT PRACTICE AUDIT GUIDE

Section II/Code of Ethics

A. Introduction

As auditors we are engaged in the formal examination and verification of the Autonetics procurement system to insure that all aspects of this system are in conformity with the Corporate Procurement Policies and that the system is effective and efficient at all times. Consequently, in our work, we may have access to confidential company records that contain information which must be safeguarded. We also will have access to the job classification and performance level of each individual buyer within the procurement function. As a result, an attitude of high standards of ethical principles have been adopted by the Management of the Quality Program Audit function. Not only do we represent the profession of auditing, but we carry with us the stewardship of high standards of conduct, honor, and character placed on us by Management in the fulfillment of meaningful auditing practices. Therefore, a Code of Ethics has been prepared to outline the standards of professional behavior for the guidance of the Quality Program Auditor.

B. Code of Ethics

The provisions of this Code of Ethics is concerned with basic auditing principles. The auditor should be cognizant that individual and independent judgment is an absolute requirement in the application of these principles. The Auditor has a responsibility to act in good faith and to maintain integrity beyond question. The Auditor will be restricted only to the limit of his/her technical skills; however, they will promote the highest auditing standards in terms of advancing the interest of Rockwell International, Inc. (The Company).

C. Articles of the Code of Ethics

1. The Auditor shall have an obligation to exercise honesty, objectivity, and diligence in the performance of his/her duties and responsibilities.
2. The Auditor shall adhere to the highest professional Code of Ethics in the review of internal controls and compliance of the MPG, AMP, AOM, and generally accepted Procurement Practices.
3. The Auditor shall conduct his/herself in accordance with professional standards for mode of dress, manner of speech and behavior considered acceptable throughout business and industry.

4. The Auditor, in holding the trust of the Company, shall exhibit loyalty in all matters pertaining to the affairs of the Company, or to whomever he/she may be rendering a service.
5. The Auditor shall not knowingly be a part of any illegal or improper activity.
6. The Auditor shall be prudent in the use of information acquired in the course of his/her duties.
7. The Auditor shall not use confidential information for any personal gain or in a manner which would be detrimental to The Company or persons with The Company.
8. The Auditor, in expressing an opinion, shall use all reasonable care to obtain sufficient factual evidence to support his/her position.
9. The Auditor shall continually strive for improvement in the proficiency and effectiveness of his/her service.

(Reprinted with the permission of Rockwell International, Inc.)

Figure 6.3: THE INSTITUTE OF INTERNAL AUDITORS, INC. CODE OF ETHICS

Introduction

Recognizing that ethics are an important consideration in the practice of internal auditing and that the moral principles followed by members of *The Institute of Internal Auditors, Inc.*, should be formalized, the Board of Directors at its regular meeting in New Orleans on December 13, 1968, received and adopted the following resolution:

WHEREAS the members of *The Institute of Internal Auditors, Inc.*, represent the profession of internal auditing; and

WHEREAS managements rely on the profession of internal auditing to assist in the fulfillment of their management stewardship; and

WHEREAS said members must maintain high standards of conduct, honor and character in order to carry on proper and meaningful internal auditing practice;

THEREFORE BE IT RESOLVED that a Code of ethics be now set forth, outlining the standards of professional behavior for the guidance of each member of *The Institute of Internal Auditors, Inc.*

In accordance with this resolution, the Board of Directors further approved of the principles set forth.

Interpretation of Principles

The provisions of this Code of Ethics cover basic principles in the various disciplines of internal auditing practice. Members shall realize that individual judgment is required in the application of these principles. They have a responsibility to conduct themselves so that their good faith and integrity should not be open to question. While having due regard for the limit of their technical skills, they will promote the highest possible internal auditing standards to the end of advancing the interest of their company or organization.

Articles

I. Members shall have an obligation to exercise honesty, objectivity, and diligence in the performance of their duties and responsibilities.

II. Members, in holding the trust of their employers, shall exhibit loyalty in all matters pertaining to the affairs of the employer or to whomever they may be rendering a service. However, members shall not knowingly be a party to any illegal or improper activity.

III. Members shall refrain from entering into any activity which may be in conflict with the interest of their employers or which would prejudice their ability to carry out objectively their duties and responsibilities.

IV. Members shall not accept a fee or a gift from an employee, a client, a customer, or a business associate of their employer without the knowledge and consent of their senior management.

V. Members shall be prudent in the use of information acquired in the course of their duties. They shall not use confidential information for any personal gain nor in a manner which would be detrimental to the welfare of their employer.

VI. Members, in expressing an opinion, shall use all reasonable care to obtain sufficient factual evidence to warrant such expression. In their reporting, members shall reveal such material facts known to them, which, if not revealed, could either distort the report of the results of operations under review or conceal unlawful practice.

VII. Members shall continually strive for improvement in the proficiency and effectiveness of their service.

VIII. Members shall abide by the bylaws and uphold the objectives of *The Institute of Internal Auditors, Inc*. In the practice of their profession, they shall be ever mindful of their obligation to maintain the high standard of competence, morality, and dignity which *The Institute of Internal Auditors, Inc*., and its members have established.

7

Product Liability and Ethical Considerations

Every year there are millions of accidents where injuries are sustained and property is damaged. Many of these events are associated with products. The U. S. Consumer Products Safety Commission maintains a system for sampling hospital emergency rooms to gather statistics on the frequency of accidents requiring emergency room treatment and the product that is associated with the incident. The products vary from the trivial to the extremely expensive and complex. Automobiles, firearms, tobacco, and alcohol are not part of this organization's responsibility and are not listed. Parts of buildings, tools, garden and sports equipment are involved in these accidents. In some instances the product itself is not to blame, but the operator of the equipment is at fault. In some instances the product or the device is at fault and performed in an unexpected and undesired manner. The National Electronic Injury Survey System (NEISS) collects this information from approximately 100 hospital emergency rooms nationwide and estimates the number of accidents per year associated with products. It is estimated that there are more than 1.4 million accidents per year associated with stairs, ramps, and landings; more than 100,000 with manual home workshop tools, more than 400,000 with drinking glasses and knives, and more than 40,000 with lamps and electrical fixtures and switches. The number of injuries associated with sports and sporting equipment is even larger.

People are injured by almost every device. They slip in tubs, are scalded by water that is too hot but that comes from a tap, and are burned when garments they are wearing are ignited by stoves. Is it the fault of the stoves, or the garments or the people or partially all three? Knives fall off tables and puncture the shoe and the foot. Is it the fault of the knife?

For every product that has been placed on the market it is probably true that someone has suffered a loss, sustained an injury, or suffered an inconvenience. The injuries suffered from some products are more serious than those caused by others. One of the products which has a long history of causing explosions is the lead-acid storage battery. This is used in millions of cars and other applications. The number of these batteries in service is extremely large. Many blow up resulting in people losing their sight. The industry has markedly improved the performance of batteries in cars and as a result the number of accidents reported to the Consumer Products Safety Commission's National Electronic Injury Surveillance System has dropped. Many of these injuries can be attributed to failure of the products due to design and manufacturing deficiencies. Are you aware that the battery in your car could explode and blind you while you were leaning over it? Is it proper that this hazard in not more vividly displayed and explained to every automobile owner and to every person who services a car?

The undesired performance of the device may be any of a variety of effects, such as overheating and causing a fire or emitting noxious fumes. Mechanical devices may perform in such a manner that they do not stop when the brake is applied or they overrun or behave erratically. Sometimes mechanical devices pinch, squeeze, cut, or reverse direction. There are unexpected and unscheduled maneuvers some machines perform. The resultant loss or injury is blamed on the machine or product.

In the United States and in many other countries the injured parties will expect manufacturers, sellers, installers, or owners of machines to compensate them for medical expenses and injuries. It is not the purpose of this discussion to delve deeper into the why and wherefore of the expectations. In the United States and other litigious areas, injured parties are very likely to ask for substantial sums from manufacturers and/or installers. In the event the suggested settlement is insufficient the injured party may go into court to try and obtain a suitable recovery. In some instances it has seemed that injured parties are asking for much more than that to which they are entitled. In other instances, you may wonder why the designers and manufacturers did not foresee the possibility of the event and why they did not build a safer product or one which was free of defects.

The term *defect* in the quality control lexicon is a nonconformity that adversely affects usage. By this statement, it includes all forms of dangerous and hazardous conditions. A nonconformity is a deviation from the intended design. Unfortunately there are designs that are improper, and hazardous, and far below state of the art. Such designs are defective and all of the units produced from that design are or may be hazardous and defective.

When someone is injured or sustains a loss caused by a defective product, has there been an ethical failure? To really come to a reasonable conclusion must ethics be redefined? If as is common practice people come to the conclusion that the quality control system failed, then the person who was responsible for maintaining the quality control system failed. Also, since we must assume that the person in charge of the quality control system was a professional and a member of the American Society for Quality Control, there was an ethical failure. The individual did not install, maintain, operate, and administer a quality control system that was effective. Of course we have made a lot of assumptions in reaching this conclusion.

It must be known to most people that each and every individual in a small or large corporation is not able to operate as an individual, and that the corporation and its employees all exert certain pressures and restrictions on each and every other member of the organization. The design, manufacturing, purchasing, sales, as well as the financial and personnel departments all exert certain pressures and restraints on what any one person can do. If it were not so, we would have total turmoil, a set of conditions that no one really desires. Nonetheless there are other professionals in the system. If they exert adverse pressure on a quality system and its members to do less than what they really should to prevent defective product from being manufactured and leaving the plant, they may also be suspected of having failed to behave in an ethical manner. In some companies there are many forces which seek to have as much product shipped as possible. There are pressures which discourage care in manufacture and assembly. Sometimes even efforts to avoid any inspection and disincentives for effort to improve safety as well as other factors may or may not come under suspicion as being ethical failures.

Hedge trimmers are not only capable of causing great injury, but they do. They have cut through power cords, and have amputated fingers and caused other injuries. If this is an argument that the product is unsafe, should it be outlawed or otherwise restricted? Is there a different standard that should be applied to the ownership of a rotary power mower, a hedge trimmer, a very sharp knife, a machine gun or an AK 47? Is there a different ethic or a different set of ethical values that should be applied to the ownership of one or the other? If so why? The lawn mower and the hedge trimmer can do serious damage, and the AK 47 can kill. Is there an argument that can be made that one has a more utilitarian purpose than the other?

Electricity can have beneficial and detrimental effects. There are deaths and injuries from contact with ordinary household power and from contact with the high-voltage lines that the utilities use to distribute power. Some lawyers have embraced the concept that electricity is a product, and that when it serves to perform some function that is not intended, like causing severe shock, burns, and/or death, the electrical supplier has a responsibility. Other lawyers have pursued the philosophy that in many of those cases where people had been injured, the utility had not taken the appropriate action that should have been pursued to provide protection, and because of this negligence, the utility has a responsibility. At other points in this book there have been presentations of actual cases.

The following instances describe real cases where there may have been ethical slips. You will have to make up your mind as to how you would assess the ethics involved.

Case 1: Injury in a power station

At the time of construction of a six building office complex, the developer was led to believe that he could install and operate his own power system thereby generating electricity more cheaply than if bought from a local utility. He planned to operate the complex and believed that the expense reduction resulting from generating his own electricity would markedly increase his profit margin. The power would be generated in one of the buildings and the main feed would be a tie line to a distribution center in an adjacent building. The distribution center consisted of a series of enclosed consoles, each of which contained a remotely activated circuit breaker and a supervisory panel. The tie line fed a bus system that was common to the distribution consoles and this in turn was connected to the tie line through a central console that contained a disconnect circuit breaker. Despite the fact that this was done in the United States, the equipment was all made in England and installed by a British firm. The firm brought a British supervising engineer to this country to manage the installation.

The firm did a complete installation and left the scene with everything operating properly. They also left a neat, clear, one-line diagram showing the path of the feeders from the generating site to the distribution site and to the six buildings and special installations. In the arrangement the power was fed into a console located in the center of the row of consoles, and the power was fed to the various sites from the consoles on each side of this connection.

As time went on the operator of the complex found that his costs were higher than anticipated and that power could be bought from the local utility for considerably less. Arrangements were finalized and the local utility connection was made to the distribution bus. The original power-generating equipment was removed and sold.

Sometime later there was a very heavy rainstorm and the basement where the distribution panel was located was flooded. Not only was power lost to the entire complex, but there was also the nasty job of pumping out the basements and cleaning and drying all of the electrical equipment. Having been in touch with the installers, the owner decided to have the supervising engineer come to the site and check to see that the equipment was all properly dry and in a condition to operate safely before power was restored.

When the engineer arrived he naturally looked for a diagram indicating where the connection between the utility and the distribution bus had been made. There was no new diagram available. Those at the site assured him that the connection to the utility had replaced the original connection to the generating equipment, and that to all intents and purposes, his original diagram represented how the bus was set up.

The engineer knew from the old diagrams that the connection had been made to the console at the center. When that breaker was open the entire bus was disconnected, and it would then be safe to enter the other consoles and check the equipment and insulators for moisture. In addition, to provide greater safety, the engineer opened the other end of this cable so that to all intents and purposes the cable feeding the distribution bus was presumably dead, and the distribution bus was dead. He also grounded the distribution bus and posted safety signs indicating that people were working on the lines so that no one would accidentally return power to the system.

With the help of a local assistant, the engineer unlocked and entered each console while the assistant held a battery-operated light behind him. The engineer entered, viewed, and felt around each insulator to check for dampness. Everything seemed in order.

The engineer had checked the insulator in the last console which was on one end of the bus and found it dry. He thought that he noticed something loose behind one of the insulators. He asked the assistant to raise the light and then proceeded to reach behind the insulator once more. There was a terrific flash. The assistant was blinded and dropped the light and it went out. The engineer had been careful, and had been reaching with one arm only. The remainder of his body was clear of all other electrical connection. The flash had burned his arm very badly but he was otherwise unhurt. He had not been looking at his arm and was not blinded. The engineer led the blinded assistant out of the basement and up a flight of stairs even though it was dark. An ambulance was called and both were taken to a local hospital.

A large hole had been burned in the engineer's arm. He was hospitalized briefly in the area and then returned to England where he was treated under the United Kingdom's health plan. He later requested payment for the injury and partial loss of power in his arm and hand.

What had happened was that the utility and the installers of the utility connection had agreed to take a slight short cut and save cable. They connected the utility's feeder to the end console and reconnected the cable that had gone to this console to the middle console which had been the original feed. No one thought to make a new diagram. When the engineer had reached behind the insulator he had contacted the live high-voltage cable which was connected to the public utility.

The organization that had hired him and their insurers refused compensation on the basis that he had exceeded his authority, which was to examine the system for moisture.

The case came to trial in a U. S. court. Do you believe that the engineer, the organization that hired him, and the insurers, all behaved in an ethical manner?

When those portions of the engineering codes that define diligence are reviewed, it is found that engineers have a responsibility for public safety and the health of others. The recognition of responsibility can be traced back to the U. S. Steamboat Code that was developed to reduce the boiler explosions and the loss of life and property that was occurring on the Mississippi River. The code was passed by Congress in 1852. The medical and engineering professionals have this open responsibility that does not always exist in a statement of the ethical behavior of all the other professions. In some instances professionals are encouraged to act in a manner that is most loyal to their employers and admonished that they must keep the secrets of their employers. In some instances these secrets are not in the interest of the general public. We even see instances in public service where underlings are criticized for not displaying more loyalty to their supervisors. This is particularly true when their supervisors have been dishonest, have accepted gifts and bribes, and have aided special interests rather than doing the best that could be done for their constituency.

An interesting statement of responsibility and how to evaluate it is contained in the second edition of the National Electrical Safety Code which defines how public utilities shall interpret this responsibility. This was published prior to 1920.

> There is no intention of requiring or even recommending more expensive construction than good practice requires and good business justifies. But it must be remembered that the public in the end pays whatever extra cost is caused by requiring safer and better construction, and hence the public may rightly require a good degree of safety in the construction...

An important part of this statement is akin to the statement that the public and general business conditions demand good quality. But in that case, as well as in the safety issue, there is no assurance that greater safety, just as better quality, must cost more. In fact the failure to provide safe conditions can readily lead to large losses when people are injured and successfully sue for large compensatory damages.

The IEEE Code of Ethics stipulates that members shall: "Accept responsibility for their actions; Undertake technological tasks and accept responsibility only if qualified by training or experience." The code also stipulates that: "Members shall, in fulfilling their responsibility, protect the safety, health and welfare of the public..."

In the same vein the ABET Code of Ethics for Engineers states that: "Engineers shall hold paramount the safety, health and welfare of the public in the performance of their professional duties."

The admonition of the ASQC's Code of Ethics is that: "Each member of the society shall uphold and advance the honor and dignity of the profession in keeping with high standards of ethical conduct." Each member is expected to: "Use his knowledge and skill for the advancement of human welfare and in promoting the safety and reliability of products for public use."

In light of these admonitions, it might be considered that engineers involved in the design, manufacture, and distribution of products that injured individuals might have performed their duties in a less than ethical manner.

Another way to look at these same questions of whether actions are ethical, is to look at some incidents that have occurred in the past and ask whether you would like to have been the designer or the manufacturer of the product or project. Therefore, you would be in a position to proudly tell your grandchildren about the activity some years later. Let us look at another case and ask the same questions.

Case 2: The Tacoma Narrows Bridge

The Tacoma Narrows Bridge which now connects the city of Tacoma to the Olympic Peninsula replaces the one which was built in 1940 and collapsed four months later in a windstorm. The original bridge was long and narrow. Due to the relative flexibility of the structure and the aerodynamics of its roadway, crosswinds would cause it to sway violently, oscillate, and go into wavelike motions. The structure immediately gained the name "Galloping Gertie." These violent oscillations caused the roadway to break up and the span to collapse. This event occurred before the era of liability claims. There were several interesting items in connection with the event. As I recall no lives were lost.

The bridge authority had insured the bridge. The agent who wrote the policy was of the belief that such insurance was unnecessary and therefore just pocketed the premium. In general, insurance on large structures and other large projects is spread around. The initial underwriter buys insurance from other underwriters to spread his or her risk, thus reducing the loss in the event of a major catastrophe. The original underwriter was therefore faced with indemnifying the bridge authority even though he really had received no premium. The theory here is that the agent acts on behalf of the insurance company and that, in effect, he had committed the company even though he had wrongfully pocketed the premium. We will not question this ethical failure.

The Whitestone Bridge linking the boroughs of Queens and The Bronx in New York City has similar characteristics. Its length to width ratio was somewhat smaller, but the galloping nature of the span was most disturbing to motorists driving across. It was feared that some day this bridge might also collapse.

George Steinman, P.E., a prominent bridge designer, provided corrective design plans for the Whitestone Bridge. He had stiffeners installed along the edge of the roadway to prevent the wavelike action and also damping plates to reduce the vibrations and oscillations. This markedly improved the stability of the structure, but on a windy day motorists can still notice some small annoying galloping activity.

Steinman is reported to have said that had he been asked he could have saved the Tacoma Narrows span. First he would have requested the bridge authority to provide a large number of heavily loaded trucks and park them on one side of the roadway. He would also have installed stiffener trusses that would in the end have prevented the wave motion. The aerodynamic surface would also have required attention. Recently bridge designers have taken care of these problems.

Was Steinman, from an ethical point of view, correct in not offering the Tacoma Narrows Bridge Authority suggestions to the effect that he was able to provide a solution to their problem? Likewise, was the authority acting ethically in not going to the foremost bridge designers of the time to request assistance in the solution to their serious problem? Was the designer of the Tacoma "Galloping Gertie" unethical in accepting a task for which he or she was not really technically equipped?

There is a host of other incidents. Each day there are thousands of accidents associated with products and services which injure people. Many of these wind up in needing the services of medical emergency personnel and result in product liability claims. Most are mundane and never reach the headlines of large city papers. Roughly 50,000 people are killed in highway vehicle accidents a year. Some of these lives might have been saved, had the federal authorities followed the same path that was followed in the design and mandating of seat belts for vehicles overseas. The U. S. authorities followed a different path [19].

There are spectacular events which are more startling and include the following, in addition to the Tacoma Narrows Bridge.

The Hood Canal Bridge failure no fatalities 1981
Tampa Bay Bridge, FL 1980 approx.
New York State Thruway Bridge 10 fatalities 1987
Mianus River Bridge on Interstate 95 on the Connecticut Thruway at midnight approx 1981
The *Boeing 737* in Hawaii (skin failure) 1988
Kansas City Grand Hyatt Skywalk collapse approx. 1980
Ford Pinto accidents and fires 1970's
Dalkon Shield problems resulting in bankruptcy of A. H. Robbins 1985
Bhopal, India accident thousands dead and injured 1984

Thalidomide birth defects 1957-61
Hartford Civic Center roof collapse no fatalities 1978
L'Ambiance Plaza collapse in Bridgeport, Connecticut 28 dead 1987
Challenger loss 1987
Titanic sinking 1912
The DC10 that crashed in Sioux City, Iowa 1989

In addition to these there are the everyday accidents that happen on the baseball field, during football, basketball, and other games, and the unending series of injuries and deaths that result from the contact of ordinary people with ordinary products. In addition there are spectacular air crashes and accidents including the one in Hawaii where the part of the upper shell of a *Boeing 737* blew away and one person disappeared. In this incident, it would seem that it might be termed a tribute to the pilots and designers of the plane that it was possible to land the craft without injury to others.

There is a large number of devices that I have known to interact adversely with people resulting in serious injury, financial loss, or death. Those discussed in this chapter include: the Mooney MU2 crash, respirator nonuse, electrocution while using a power saw, electrocution of a gardener. The case of the capacitor-computer interaction resulting in substantial loss to the computer manufacturer is discussed in Chapter 13.

In each of these events you should ask: Would I have been distressed if the event involved a product for which I had a responsibility? If I had been involved with any of these products, projects, or services, would I like to make a record of my involvement and pass it along to my family?

The Hood Canal Bridge, mentioned in Chapter 6, is a floating bridge providing a main highway link between Seattle, Washington and the Pacific Coast. The Olympic Park and large recreational areas exist on the Olympic Peninsula. The Hood Canal is not a canal, but really a sound. The floating bridge is supported on concrete platforms that were designed to withstand ordinary storms, but not the storms that occur, it is estimated, once in 100 years. In 1981 such a storm occurred and several of the floats were lost and the bridge put out of operation for two or three years. Was it ethical for the state of Washington to build and rebuild such a bridge? In all fairness it must be said that during the time the bridge was being rebuilt a ferry service was placed in operation. On summer nights the ferry lines were extremely long.

Another bridge was in the Tampa-St. Petersburg area in Florida. A ship hit a bridge pier and wrecked it. A similar situation occurred near Cape May, New Jersey where the Brunnel (bridge-tunnel) crossing was put out of service by a ship. Were the bridge designers ethical in not taking into consideration the fact that since there were ships passing in close proximity to the supports, a ship might hit and destroy one of them?

A bridge on the New York State Thruway, near Syracuse, was washed out by a swollen creek and heavy flooding in 1987. Ten people died. Traffic was bypassed for almost two years, and there was a lengthy investigation. The "as built" drawings and the inspections of the bridge had failed to disclose that the supports for the bridge did not include the heavy stone protection (rickrack) that should have been applied around the underwater portions of the supports to prevent them from being washed away. Additionally, wire rope screens are often installed to protect the rickrack. Were the designers, builders, inspectors, and operators guilty of unethical conduct, and if so how?

The Mianus River Bridge on Interstate 95 in Connecticut collapsed at midnight causing only a few deaths. I had a young visitor who had driven over it about 12 hours before the accident. He was greatly upset. In this case only the fact that the event occurred at midnight prevented the injuries and deaths from being in the hundreds. This bridge span was relatively short. It was held in place by resting on pins which rested on a partial support on ground piers. The pins shifted and worked their way out, or according to other reports, rusted out, allowing the bridge to collapse. There had been no bolt or weld to prevent the pins from shifting, no welded brackets to support the span in the event the pins failed, and no way an inspection of the pins could be performed. The State of New York had rejected and prohibited this design. Were the engineers who favored this design, the Connecticut engineers, or the inspectors who failed to discover the conditions of the support pins or report that they could not be inspected, guilty of ethical failure?

In one case in which I was involved, an airline pilot, after completing a long flight, rented a small twin engine Mooney MU2 to fly from an East Coast terminal to a small airport in a southern state. He was in contact with the tower of a field near a beacon where he was scheduled to make a procedural turn according to his filed flight plan. At this beacon, which he was approaching from the north, he was supposed to turn west. At approximately the time he passed over the beacon, communication between the pilot and the tower ceased. There was no apparent reason. The pilot did not make his turn. Communication was never reestablished with the pilot. Approximately 150 miles south of the tower the plane was observed flying across a highway at a steep angle. It crashed into the ground at this angle.

A passing motorist who was also a licensed pilot witnessed the crash. He testified that the airplane was trailing a smoke stream. The fire that occurred on the ground was so intense that the cause of the pilot's death was never identified. He might have died in the crash or been dead from natural or other causes before the crash. He might also have been in a condition which rendered him unable to control the plane.

The crash of the plane resulted in a product liability suit. The pilot had no family and no heirs who were in a position to file suit for wrongful death. The owners of the aircraft and the insurers of the plane were interested in recovering as much as they could. They charged that one or more components were defective, and they were entitled to recover the value of the plane from the manufacturers of these components. The National Transportation Safety Board (NTSB) was also interested in discovering the cause of the crash as a quality control procedure that might result in the prevention of other crashes of similar planes.

In an effort to recover the cost of the aircraft several of the suppliers of electrical and mechanical equipment on the plane were sued by the owners. There were many theories put forth to explain the crash. The pilot might have been asleep, or have suffered a seizure. Perhaps the communication equipment had been disabled by a short circuit and fire in the electrical system, or an electrical fire had created smoke and fumes which caused the pilot to become comatose.

None of the lawsuits ever went to trial. Offers by several defendants to settle the case for sums which they judged might be less than the cost of defending the case in court were accepted by the plaintiffs. Certainly this was more than they would get if they lost the case, and the defendants were paying less than they would if they lost the case in a jury trial. The question of ethical practice for the lay person is one in which he or she might decide whether it was ethical to file a suit with so little chance of winning, but a big chance of forcing a settlement. The plaintiffs, in these cases were not the owners of the plane, but insurance companies that had settled with the owner. Many large product liability cases similar to this are really between insurance companies rather than between a true injured plaintiff and defendants.

In another small aircraft disaster four people were killed when the plane caught fire shortly after takeoff. The NTSB determination concluded that there had been an explosion and fire in a compartment toward the rear of the aircraft. This compartment contained tubing and connections through which jet fuel flowed under positive pressure, a direct current rotating machine (motor or generator) with an open commutator, and batteries. The batteries were supposedly sealed and connected to a set of manifolds which circulated air over the internal battery vents, taking air in through a forward-facing vent and exhausting air and gases through a rearward-facing vent on the opposite side of the aircraft. This arrangement effectively removed the explosive hydrogen and oxygen mixture from the aircraft, provided it was properly installed and connected.

There were several theories presented during the preparation for trial. It had been determined that the fuel system had not been assembled properly. As a result fuel might have been sprayed in a fine mist into the compartment and been ignited by the sparks from the commutator. This would have resulted in a fuel-fed fire. It was also suggested that the battery fumes might have been leaking into the compartment due to improper assembly of the vent tubes. The battery fumes contain hydrogen which some suggested might have been ignited by the commutator in the bottom of the rear compartment. Both of these theories are based on the assumption that the last service effort in the rear compartment had been faulty. There were other possibilities, but all included the interaction between the open commutator in the compartment and the possible presence of an ignitable gas or vapor.

Ethical questions surrounding the crash include: Should the plane's designers, outfitters, or other engineering personnel have made a system failure analysis which included the questioning of the open commutator in the same compartment where explosive and flammable fumes and or gases might be present in the event of a malfunction? Did the NTSB make similar analyses? Were the people who serviced the plane certified in any manner? Was an inspection made after the last servicing by someone other than the individual who performed the service? Can you identify sections of any of the codes of ethics that apply to any of these operations?

In the actual case, the battery manufacturer was successful in establishing that the battery showed no signs of an internal explosion and that the damage to the battery was the result of the crash. Therefore he suffered no loss beyond the cost of defending against the plaintiff.

As noted earlier, batteries and other devices do, on occasion, behave in a manner that results in injury. The Consumer Product Safety Commission's NEISS estimates that in 1988 more than 10,000 hospital emergency room visits were required due to injuries related to batteries. Many of these were the result of exploding automotive batteries.

In a southern state several companies who made respirators used in guarding against silicosis were defending themselves against a plaintiff who was in the last stages of the disease. The suit had been filed to gain money for his wife and children. The individual was an inmate of a federal prison. Many years ago he had worked in a location where respirators should have been used but were neither furnished nor used. These manufacturers had been in the business of supplying these devices to industry at that time. Would you comment on the apparent ethics of this case? As a manufacturer how would you feel?

Ethics, which for the purpose of this book, are the engineers' duties to conform to a code of ethics of the organization to which they belong, or as stipulated by the licensing code of the state(s) where they maintain registration.

There are many instances where engineers are faced with decisions wherein they must take or recommend one of several courses. The decision to follow one path or another is generally made on the basis of which course is more economical, or more beneficial to the organization for which they are working. The question of ethics is not usually recognized. The decisions which offer cost savings or expediency are often difficult to resist. In other instances precepts of the engineer's ethical code are ignored or contravened, more often than not, I suspect, due to failure to recognize that ethical considerations should take precedence over other considerations. Have you looked at your society's Code of Ethics recently? Do you have it within easy reach of your desk?

From an engineer's viewpoint ethics might be said to be ignored in any restrictive legal effort to minimize the probability of a plaintiff obtaining a competent engineering expert to aid in the preparation of his or her case. One of these restrictions has taken the form of rules or procedures which attempt to prevent, or do prevent, the court from recognizing and allowing an engineer to testify as an expert if he or she has not designed and manufactured a device similar to the one involved in the incident being litigated. Another rule attempts to disqualify a plaintiff's expert if litigation fees are more than a small percentage of the expert's annual income.

Consideration of several examples from industrial and forensic practice follow. The examples illustrate how ethical considerations were neglected or perhaps not even recognized by engineers, as the engineering aspects of the problem were reviewed. As a result of ethical and legal oversights, people were injured or losses incurred.

Two large corporations had divisions which had a partnership in delivering a large quantity of material to the U. S. Department of Defense. The contract ran out and the two vice presidents of the divisions were each left with a large plant and a large labor force which they would have to lay off. The situation was so bad that they both feared that their divisions would be disbanded and, of course, the ultimate would be that they might be looking for work elsewhere. This incident predated the computer revolution.

During a visit to the other's plant, the vice president of A spotted a device which resembled a video terminal and could be used to transfer data between sites. He talked to his staff about this product. They were of the opinion that given this device, they might be able to sell as many as several hundred thousand, particularly since it could be produced at an attractive price. An agreement between A and B was reached, and B agreed to complete work on this product, deliver the units, and service the factory returns. A would sell the unit and provide field service on the product. This was going to be a gold mine for both. It would keep the plants operating and the divisions in business. Some months later the product was not selling, and field service seemed greater than had been anticipated. The glow of the pot of gold at the end of the rainbow had disappeared. A canceled the contract and sued B for some $20 million for loss of profit.

The main claim was that the product was so unreliable and needed so much service—far beyond that envisioned or agreed to in the contract—that the cause of A's failure to sell millions of dollars of this product was its unreliability. It was claimed that B's failure to perform and deliver a reliable product in accordance with the terms of the contract was responsible for A's loss, and for A's failure to make millions of dollars in profits.

There were at least four competent engineering experts on this case, and they were to work as a team. The defense of this suit required a careful review of each statement. At the outset the attorneys who interviewed the group wanted to have them work as a team and be able to agree to all statements and vouch for their authenticity. At this time the attorneys seemed most interested in how to define reliability. The engineers voiced dissenting opinions: regardless of the way reliability was defined, what were the facts? How long did these units actually stay in service? Was the need for service a major factor in the return of the units after a trial in the field?

It was extremely difficult to obtain accurate figures on field service and replacement requirements from either side. It is often difficult to get data from the opponent, but in this case it was even more difficult to get real data from the company being defended.

Apparently, the reliability of the product was to be in excess of 3,000 hours between requirements for service. If used 10 hours a day, this would be the equivalent of 300 days between service requirements. Today, this seems like a short life. The figures obtained from the plaintiff indicated that he was complaining that the device needed service at intervals of once a week. There had to be a means of determining what had happened, and whether the time between service needs was 30, 1,000, 3,000 or more hours, on a realistic basis.

If the data indicated too low a life (markedly less than 3,000 hours) the plaintiff was in trouble. If the life was marginal the case could be defended but if it grossly exceeded 3,000 hours, the defendant had a wonderful case.

At one time an investigator for the attorney went to an area where an organization had contracted to buy 100 of the devices. The vast majority, according to the plaintiff, had been returned because of service difficulties. The investigator found that there were only three of the units now in the hands of the user. The balance had been returned, and were not in service! The investigator was sent back to find out whether the organization was using another similar device or a competitor's unit. By this time, several years had elapsed, and there were now several competitors.

It was found that the organization was doing all it needed with the three units it had, and that there was no need for the other 97. The reduction of the number of units in service was due to lack of need, and not failure to provide units with good service experience.

The plaintiff's records indicated that all the returned units had failed in service. Yet there was no indication that the units had been serviced or repaired. There was evidence in the plaintiff's record that many of the devices were removed from one customer's location and immediately installed at another customer's site.

The first factual data was obtained from the complete service experience of the plaintiff. It indicated a mean life close to 3,000 hours. It could not be said with certainty that the mean life was over or under, but it was much greater than the claimed 30- to 300-hour breakdown rate.

Why was the defendant withholding data from his attorneys? The attorneys finally got a sample set of data. It showed something in the order of 300 hours between service calls, and began to explain the unwillingness to provide data. Checking the serial numbers of the units included in the plaintiff's data, it was found that the data from the plaintiff covered only those sets which had the most frequent service requirements. The remainder of the sets taken as a group had a service requirement of much less than one call per 3,000 hours.

Why were the attorneys given that misleading data? The plaintiff claimed that he understood that to be what was wanted. The plaintiff's representative seemed to have no concept that a requirement of 3,000 hours between service calls was an average, and did not refer to each and every unit. Apparently the internal communication at the plaintiff's offices was less than adequate. They had competent engineers, but did not use them.

This case never went to trial and was finally settled for a nominal value. The experts' fees ran well over $100,000, and the legal fees on each side probably exceeded $1 million.

The people working on the defense were engineers and statisticians, all skilled in electronic reliability. Not one of them had ever been involved in designing or manufacturing a device of the type being litigated. The calculation of the reliability of a product of this type could be well established by reference to the literature. The people who were responsible for the design and manufacture of the product were, in my opinion, competent to do what the expert team did. They had more familiarity with the product. There was apparently much needed in the way of internal communication.

Aspects of the case that worked to the advantage of the defendant were:

- a close working relationship among the members of the defense team;
- regular reviews of progress, findings, and questions;
- an openhandedness with documents on the part of the attorneys,who provided all the engineers with complete copies of reports, depositions, and abstracts;
- findings of one expert were circulated to the others;
- what each expert or team was doing and was to do was openly discussed at conferences.

The case was large enough to provide sufficient funds to easily support such activity, yet the organized approach probably saved effort and helped sharpen the concept of the attorneys and improved the efficiency of engineering investigations.

Were there ethical shortcomings on the part of engineers who worked for either A or for B on this project?

It is difficult to understand how competent engineers for both plaintiff and defendant could be ignorant of, or could ignore, standard methods of calculating and demonstrating reliability. Perhaps this was an instance in which engineers had failed to maintain a familiarity with the art in which they were engaged. Perhaps sales and contract people operated without engineering guidance.

This was an unfortunate contract for both sides pressured by the executives who were pressured by the lack of work for their division. Executives, procurement, and contract personnel appeared not to have used the services of engineers available in their systems. In addition suggestions made by the engineers that were consulted seemed to have been ignored. The engineers seemed to have made too little effort to convince the executives that they were pursuing the wrong track. Would this be considered an ethical failure? You ought to examine some of the codes of ethics to see how they were violated in this case.

Another more serious case, involving a power saw, had many curious twists, including disappearance from the public prosecutor's office of the evidence.

A young man who was quite a clown obtained a summer job at a site where several houses were being erected. He had his own power saw and brought it to the site. One evening he replaced the power cord. Being ignorant of the proper procedure, he connected the wires so that the exterior metal housing, which should have been connected to the ground plug of the three-wire cord, was connected to one side of the line. The next day, after a rainy night, he plugged the saw into the end of a two-wire extension cord. When he picked up the saw, he began to madly dance around, much to the amusement of the others on the site, who did not realize that he was being electrocuted. Immediately after the accident, the connection to the electrical supply system was found to have no ground fault circuit interrupter.

A ground fault circuit interrupter (GFCI) is a device required in specific locations by the National Electrical Code. It is widely used in bathroom circuits, garage, and outdoor circuits to prevent serious shock and/or electrocution. It was required by law at the building site. It senses when current that goes out through one of the two feed lines does not return through the other. When someone receives an electrical shock, the electricity that goes out on one wire is returned to the system through a ground that is not part of the system. When a hot line is touched, current passes through the body or part of it to the ground and shocks the individual. Had the ground fault circuit interrupter been present, the young man would probably have felt nothing, because the line would have been disconnected in a fraction of a second, in fact, so rapidly as to prevent injury.

A feature of these devices is that they have two buttons, one of which is a test button. When the test button is pressed, a small current passes between the hot line and the ground. This is the same type of event against which this device is intended to protect. The device trips, and the reset button must then be pressed.

In this case, a Professional Engineer who filed a report on the accident for the contractor stated that the presence or absence of a ground fault circuit interrupter was of no consequence, and would not have prevented the accident. He stated that the only function of these devices is to disconnect the power when the test button is pressed. A report of this nature is bound to reflect adversely on all engineers. It makes lawyers and others involved in the case think that engineers are stupid. The individual who wrote the report was either incompetent or trying his best to confuse the adversaries, as well as his client. He was not being factual.

A report devoid of truth and so ignorantly stated is hard to imagine. Yet it was produced. The National Electrical Code, (NEC), which is invoked in most areas of the country, requires these ground fault circuit interrupters. These recommendations are adopted into the laws of many municipalities, communities, and states. How could this engineer have imagined that the ground fault circuit interrupter was a useless device?

A plaintiff's engineer fully investigated the incident and documented the performance of GFCI, why it was required, and how it would have prevented the death. The victim's estate prevailed.

The engineer issuing the report on the GFCI, stating that the device would have had no effect on the accident, may have violated the code of ethics of the Institute of Electrical and Electronic Engineers in one or more respects. Presuming he was trying to be honest, consider the following hypotheses:

- he failed to maintain his professional skills at state of the art;
- he undertook a technological task, and accepted responsibility, in an area where he was not qualified by knowledge, training, or experience;
- he did not "act as a faithful agent" for his employers.

In another case, the attorney advised the engineer to lie—tell of imaginary experiences. No one will know, he claimed. Such action might convince a jury of an individual's competence if he was to say he had designed special steering devices for an imaginary company. It might be hard to verify and prove during a short trial that the statement was untrue. What about the next time? Would that attorney trust the engineer? Will a future opposition attorney read the transcript, find the untruth, and confront the engineer? Ethically it is wrong to lie. In addition, there is no one with a good enough memory to support a lie—even those skilled in politics, where truths, half-truths, and untruths seem to flourish. Liars do not prevail for long.

The engineer has many techniques available for investigating cause-and-effect relationships. The effective use of some of these procedures is a skill an engineer brings to bear in finding the cause of the incident, the failure mode, the offending element, or party in a product liability case.

Both plaintiffs and defendants are faced with the same problem—finding just how the accident happened. Each side in a lawsuit wants to develop a clear understanding of what occurred and who was responsible. There is often a failure or an item which triggers the event. It is not always simple to identify the item. Furthermore, the failure may be due, not to an improper design or manufacture of the item, but to the way the item was applied or used. The link that failed, or contributed to the injury, even though working as planned, may have been of ideal design, but misapplied and/or improperly installed. There may have been a failure due to ignorance, which might have included failure to warn. Failures and accidents can also be the result of a number of contributing factors which could, and should have been foreseeable. The cause-and-effect chart, or a fishbone diagram, Failure Mode or Effects Analysis, Sneak Circuit Analysis, Hazard Analysis, or Failure Rate analysis may be useful.

Each of these has its advantages. All are helpful in making it possible to relate cause-and-effect relationships and identify methods by which the event could have been inhibited or mitigated.

The most important result of making a structured analysis is that it can uncover unsuspected contributory causes which can be analyzed and which may explain the true faults and causes. Discussions between engineers, between engineers and lawyers, using and explaining the technique(s) employed by the engineers, help define an effective defense or plaintiff direction. These devices also help explore the path of the devil's advocate—the claim the opposition may make. They may help to avoid surprises during a deposition or trial.

Perhaps the poorest service to the client, and one which is ethically questionable, is the failure of a member of the lawyer/engineer team to disclose all information to the other. This is typified by the lawyer furnishing only three pages from a 100-page deposition, and the engineer failing to disclose to the lawyer how he or she came to the decision that supports a conclusion or opinion favorable to the client.

The ethical considerations here may be covered in the IEEE Code of Ethics by Articles I and II. Can you identify other ethical failure references that are pertinent to some of these events?

Another case involved a gardener, who was cleaning out leaves and debris in April. He walked into a downed high-voltage wire resting on some brush. It hit him in the chest, and he was electrocuted. The utility disclaimed any responsibility, due to the fact, they stated, that the only way they had of knowing that a line was down, was through a report from a customer of an outage, or from a passerby who spotted the line.

This high-voltage line ran through the backyards of a community. It had parted, and the live end rested about chest high on a bush. It so happened that this break occurred near the end of the line. The only need for this line was a transformer that fed an air conditioner in the last house.

The break occurred sometime after the last use of the air conditioner in the fall. It was impossible to determine whether the wire had been down for months or hours. It really did not make any difference.

OSHA visited the scene, and concurred in the opinion that the utility had no way of knowing that the line was down. The statement was true. It was also true that there was no specific section of the National Electrical Safety Code that states what the utility shall provide in such a circumstance. On the other hand, there is a statement that the utility shall do everything within economic reason "to provide for the safety of the public and the employees." The economic stipulation is based on what is reasonable. The code states "all electrical supply lines and equipment shall be of suitable design and construction for the service and conditions under which they are to be operated, and all lines shall be so installed and maintained as to reduce the life hazard as far as practical."

The utility's engineers maintained that there was no economical basis for expensive sensors and breakers to prevent this type of occurrence. They did not think of the simple ways to provide protection, but rather set forth expensive and highly esoteric devices that—they rightly pointed out—were not economical. In connection with this case, a prominent consulting firm prepared a document for the utility which, in essence, indicated that the utility had done all that it really needed to do to protect the public. The report was typical of several similar reports. There was nothing in the report that was untrue. There were excerpts from the National Electrical Safety Code that indicated, or were intended to indicate, that the utility had fulfilled all of its responsibilities. The excerpts were carefully chosen to show the utility in good light. The statements were true, but they were not the whole truth. They failed to include the other half of some paragraphs, and some sections of the code.

The consulting firm had skillfully chosen a sentence, a part of a paragraph and other excerpts to prove the case. A sentence-by-sentence or paragraph-by-paragraph rebuttal, showing the other parts of the excerpt that had been omitted in order to show the utility in a good light, helped resolve the case. Considering the earlier reference to the National Electrical Safety Code are the ethics of the various parties in this case above reproach?

There are several by-products of a product liability matter or investigation. One would hope that the plaintiff and/or defendant might learn a lesson from the incident. The injured party might learn how to avoid injuries or losses, and the manufacturer or service supplier might learn a technique that would minimize injury or make injury of that type less likely to occur or to occur with less frequency. In the case of the utility, I wonder if the utility or other utilities learned of the exposure and installed some inexpensive warning device to prevent their being ignorant of a downed line when the line was lightly used?

In some instances companies have had inspections made by outside organizations or individuals, to try to identify areas of exposure and suggest corrective action. A lawsuit ought to have a similar impact on an organization. Is it ethical not to inform the engineering staff of a utility of the suggested correction? Is it ethical to ignore economical procedures which might save lives?

The science and art of statistics teaches that events which occur infrequently can occur. When such an event occurs repeatedly at the same site, or with the same equipment, or to the same person, it is unlikely that this occurrence is due to chance. There may be an identifiable cause. I am firmly of the belief that when something untoward or unusual happens to me accidentally, it is a random occurrence. When this type of event happens not once but twice or three times, it is a pattern, and indicative of something that needs attention or correction. It is not to be ignored. When I am shortchanged at a toll station, it may be an accident. When this happens at the same toll station, in the same manner, a second time in short order, it is a pattern that may warrant investigation by the toll authority.

The record-keeping of the warranty service performed by an organization which repairs its own equipment offers an opportunity for the identification of repetitive breakdowns that require attention. This may be the warranty service's major value.

The records may show a random pattern, or a few very important failure modes, which can be corrected. Similarly, complaints relating to losses and injuries which have occurred and which are associated with a product should be analyzed and reviewed to find patterns and areas where corrections and improvements can be made. Some companies have lists that show hundreds of injuries associated with one product. Some of the records from the Consumer Products Safety Commission National Electronic Injury Surveillance System (NEISS) files show that there are certain groups of products which, year after year, are responsible for thousands of serious injuries. Certainly the companies making these products are well aware of the problem, and must set aside substantial sums to take care of these losses. Despite the repeated losses, some companies seem to learn little and appear to make no effort to reduce the frequency and the severity of such injuries. Instead they will emphasize that the reports include an injury associated with product where the individual slipped and fell and hit his or her head on the product. This reference is, of course, an effort to minimize the belief in the system. Such events occur and add to the listed injuries related to that product.

As members of the quality profession we are interested in doing our part to assure that people are not injured. This is not only acting as a good citizen, but also protecting our company and our job. What should we really be doing to optimize the organization's safety, and reduce the probability and frequency of incidents which result in injury and loss?

We also have another responsibility as a member of an organization and that is to see that the organization is profitable and continues to be a good place to work and is able to deliver product that is acceptable to customers. What is a reasonable quality level? How can this quality level be achieved? Is it part of the job of the quality engineer to choose a proper level and to assure that it is achieved? If nails were delivered to a building site it would not be catastrophic to have one in 200 of these nails defective in a manner that prevented it being used. On the other hand if one in 200 of these nails was defective in a manner that caused it to fly out in all directions when it was hit with a hammer, then there would be a greater chance of injury. Obviously one failure mode and/or deficiency is more undesirable than another. Have you investigated these failure modes and the economy of delivery? People are now able to purchase product of various types that have failures to conform in the range of less than 10 parts per million. Does your product need to be of that quality? There can be no question that the fewer nonconforming units per million, the fewer the complaints, and the less the likelihood of damage, provided the design and the acceptance criteria are correct. Sometimes it is costly to get down to a few parts per million. Most often it is not. In fact the better the quality the lower may be the cost. Have you as a quality engineer or executive carefully and correctly analyzed these costs and chosen a path that leads your company in the direction of better quality? Should there be sorting or sampling?

Does your organization effectively use good statistical process control? Are the products improved because of this effort? If the products are not improved, what should be done to begin to see the beneficial effects from these efforts?

How does the product you ship compare with the product of other companies? Is the purity, quality, distributional width of characteristics substantially different, or better?

Years ago the maker of an industrial bolt developed a device that he believed was substantially better than many of the bolts provided for that purpose. His opinion was confirmed by the business he developed. Rather than being happy with the business the manufacturer continually invested in more accurate machinery and improved methods for manufacturing the bolt. The improved methods allowed him to produce a bolt that was more accurate and more uniform and stronger than much of his competition and his business thrived. As it improved he continued to improve his manufacturing equipment and his processes, and the quality of the product was respected far and wide. This striving for greater uniformity and better quality resulted in the cost of the product either decreasing or decreasing in comparison with others, making the organization successful. Can you see how your organization could follow a similar procedure and improve its economic position? You need not ask whether this was an ethical procedure, but you might ask why is it that more U. S. industrial organizations have not behaved similarly. Can you see the advantage of the long-term versus the short-term profit picture in this pattern?

Product-related incidents which result in losses and injuries should be analyzed and an effort made to reduce their frequency and severity. Economics as well as ethics demand that this be done.

A company should not only maintain a list of accidents, incidents, products involved, and causes, but it should also make regular reviews of these records and take appropriate corrective action. Which of the following would you include in the activity? The company should:

- regularly review the category of incidents;
- regularly review how many incidents occur in each category;
- maintain accurate records of who is injured and in what type of accident, and how to warn and inhibit these events;
- regularly review the quality and longevity of product in the field and of warranty activity;
- use a reliability review procedure;
- use aggressive quality systems;
- use standards;
- use design reviews;
- regularly trace all causes and determine how to reduce the frequency and severity of accidents.

Do ethical principles demand that only the most efficient of these methods be used in attempts to reduce injuries and losses? Are other actions dictated when ethical standards are considered?

8

Ethics and Free Speech

In November 1987, the IEEE Directors approved an addition to the Code of Ethics which states that a member shall make no statement that he or she "knows to be false concerning the IEEE, or the qualifications, integrity, or professional reputation, or employment of another member or employee." A second portion states that a member shall "neither injure, nor attempt to injure maliciously or falsely, the professional reputation or employment of another member or employee."

Members of the IEEE voiced opposition to these additions; claiming that these changes have a tendency to abridge the rights of free speech. The opposition is led by the Society on Social Implications of Technology or SSIT. The charges purport that the additions were adopted improperly, not discussed with the society's members, nor opened for review. Furthermore, SSIT claims that the additions were, one time, 50 years ago, part of some engineering society's codes and that they were abused.

One instance of the abuse of this part of a code occurred in the ASCE where a member was expelled for charging that there were corrupt practices involved in building a dam for the city of Los Angeles. Later the charges were proven to be true, and an engineer was found guilty of engaging in corrupt practice.

The IEEE Board assessed the complaint and claimed that the present IEEE policy and procedure manual effectively forecloses the recurrence of such an event.

While it seems that knowingly making false and malicious statements about others is not a practice to be encouraged, we might question whether such a statement belongs in a code of ethics. Is it proper to restrict and prohibit statements concerning members, employees, or the society (IEEE)? If such a restriction is adopted, should it also apply to nonmembers of the society? Are these restrictions properly covered elsewhere? Is it wrong to make such statements or are there situations where such statements, even though known to be false, are appropriate?

The Institute, a publication of the IEEE discusses such problems and gives limited space to comments by members who wish to air their views. It is not every group that can afford the money to do this, but in the case of the IEEE, which is the largest of the engineering societies, it is apparently not only possible, but encouraged. There is a vocal membership and there is dissatisfaction among some members as to the policy of the society.

At one time there was a member who violently objected to the publication of so many papers by foreign nationals. He claimed that this reduced the space available for publication by U. S. nationals. It would be interesting to know how many read material that is published in journals of other countries. The U. S. journals, with few exceptions, contain material that is written in Canada and sometimes in Mexico. Some of the journals devote space to activities in the Far East, but there is a dearth of material from communist countries. There must be some good work going on in other areas of the world. Access to some of it would permit us to get a picture of other markets and what others are doing. Is it ethical not to provide access to this information?

Professional Ethics, a newsletter of the American Association for the Advancement of Science, Committee on Scientific Freedom and Responsibility, Professional Society Ethics Group, is just one of a number of newsletters that has begun to take an interest in matters relating to ethics. Is the increased interest in the press and in the scientific community a sign that there is greater interest in being ethical? Or is it a sign that many are becoming concerned with the large number of reports of behavior and actions which if not unethical, are so marginal that they arouse interest in what is correct and what is really criminal? What is your opinion?

The newsletter editor, Mark S. Frankel, comments that there are interests in revising or codifying the standards of conduct of members of the scientific community and that "these initiatives reflect a belief that serious attention to professional ethics concerns and issues can increase the likelihood that ethical problems will be better understood, more carefully considered, and rendered more tractable" [20].

Another newsletter article is entitled "New Measures to Control Misconduct in Medicine." One called "Looking Back" discusses a workshop on scientific fraud and misconduct held in September 1987, under the auspices of the National Conference of Lawyers and Scientists. This AAAS-ABA conference was supported in part by a grant from the Sloan Foundation. Another article expressed the well-known opinion that a consultant should only perform or contract to perform services within his or her area of competence.

The second issue of the Professional Ethics report commented on Senate Bill S.2095, the Uniform Health and Safety Whistleblowers Protection Act. This legislation was introduced to help strengthen the protection available to private employees against reprisals for disclosing information on unlawful or hazardous activities and to help further protect public health and safety.

During June 1990 the Supreme Court of the United States ruled (9-0) that workers in the nuclear industry who allege that their employers fired them for pointing out safety problems may sue for damages under state laws.

This AAAS issue also discusses in "Professionalism and the Society of Professional Archaeologist," by Edward B. Jelkes, the codes of ethics of archaeologists as published by the Society of Professional Archaeologists (SOPA). Archaeologists are eligible for certification and membership in this society if they meet the qualifications of education and experience as set forth by SOPA. The society has also developed procedures for investigation of reported and alleged violations of the code of ethics. Archaeologists have a responsibility to protect and report on their investigations and to preserve material in a manner consistent with the best interests of society.

Since 1976 it was reported that SOPA has investigated 30 to 40 cases of alleged violations of the code of standards. One member has been expelled and several have been admonished.

The AAAS also reprinted a situation raised by Arthur E. Schwartz in his Public Law and Public Policy column. The question is from the September 1987 *The Chemist.* The *Professional Ethics Report* asks readers to respond to the following situation.

> Chemist John Jones specializes in forensics. He is retained by an attorney for a plaintiff in a case involving the unsafe handling of chemicals which resulted in personal injury. Jones is asked to furnish a safety analysis of the disputed situation and to be prepared to testify as to his findings in court. He accepts the assignment but subsequently determines that he cannot in good conscience render a report favorable to the plaintiff and indeed, the plaintiff, not the defendant may be at fault. Thereupon, the attorney terminates Jones' services and his fee is fully paid.
>
> Subsequently, the attorney for the defendant, having learned of these circumstances seeks to retain Jones to use his report or prepare a similar safety analysis for his client. Is it ethical for Jones to provide his services to the defendant's attorney? [21]

This is a strange situation. You may question the ethical practices of the defendant's attorney. If it is unethical for Jones, then would it be unethical for the attorney and vice versa, or is there another interpretation?

There are other situations in which Jones might have found that the entire case of the plaintiff is a fraud and that the claims for loss and injury have been concocted out of thin air or were the result of some other situation. Is there any action of Jones that is inhibited by law or ethics? May he join the defendant or should he go to the court or some other authority? He certainly does not want to be a party to a fraudulent action. What other options are open to Jones?

Another question arises when an attorney discusses a proposed case with Jones. Is there a minimum disclosure that an attorney may make that will inhibit Jones from going to the opposition on the basis that he has learned too much from the first attorney? If the attorney represents a man who was injured by matches and wants to retain Jones, is this sufficient to inhibit Jones? Suppose the attorney goes further and describes details of the case and the theory that he or she wishes to pursue and discusses this with Jones but no agreement is reached? Should this or does this inhibit Jones from working for the defendant? If the attorney calls Jones and says she represents a man in Orange County who was burned by matches and needs expert advice, and Jones replies that he has a case in that county and he wishes the attorney to identify the case and it is the one in which Jones is active, does the release of the information represent a breach of ethics? What should Jones do now? There are ethical as well as legal interpretations of this problem. The legal interpretations may not be the same in all localities. Attorneys may ask Jones whether he testifies for defendants and/or for plaintiffs. The question is supposed to elicit an answer that may indicate a bias on the part of the witness. In some instances the defense may spend a great deal more money than the plaintiff. In other instances the plaintiff is the big spender. It is nice from the expert's viewpoint to be on the side of the party with more money, since they may be more liberal with their time allowances. Ethically an expert ought to spend as much time on the matter as is legitimate and no more. On the other hand, there are situations where the cost of the suit can overbalance the losses.

An expert's duty is to tell the truth. The statement was made elsewhere that in most instances there is little likelihood of finding that all the evidence is in favor of one party.

What happens when an expert is a specialist in a field? She has developed an expertise in the subject of matches and matchbooks. She has had many cases in which she has appeared for plaintiffs who have had injuries or losses as a result of a fire started by a match. One day she is called to testify as an expert for a foreign company that manufactures matchbooks. If she testifies or renders a written report, will the attorneys in the next case quote the individual's own words in challenging her on some of her opinions? Is it wise for the expert to accept the case from a defendant? Is this a situation where ethical considerations are at issue? Is the situation different if the expert is still retained by some plaintiffs in cases involving the same product?

I have been an expert in several cases where I was not the sole authority. This may pose problems. In one case a tree fell and an arborist was called. In another, one of the experts rendered a report which contained factual engineering errors. What should I have done?

In some situations attorneys attempting to present situations in their most favorable manner hire several experts to work with them and their clients to develop a set of hypotheses and reconstruct the incident. This is probably more important to a plaintiff's attorney than to a defendant's attorney, particularly when the defendant is a large corporation. The corporate engineers can be consulted to appraise the opinion of the outside experts. Attorneys without other backing need to reassure themselves that the experts are correct in their assumptions and conclusions. Attorneys must be assured that the other side will not prove the presentation to be grossly in error. The experts may work as a team or may be employed independently.

As a team the group can agree on a presentation and a factual and scientific explanation. Therefore, the experts may place checks on each other. The attorneys are free to add or dismiss experts as they wish, within the scope of the authority granted by the client.

In one case the plaintiff's attorney was cross-examining the defendant's expert. The attorney read a portion of the expert's opinion in a similar case where the expert had testified for the plaintiff. The attorney then asked the witness to explain why he had come to the exact opposite conclusion in the previous case, despite the fact that there appeared to be identical circumstances.

The witness was hard pressed to give a good answer. The judge commented, outside of the hearing of the jurors, that if the plaintiff had reached the expert before the defense, the expert might well have testified for the plaintiff and reached an opinion similar to that given in the other case.

Is it ethical for a witness to reach two dissimilar opinions when the cases are identical? Is it ethical for a judge to voice an opinion such as that noted in the presence of a jury? Is it wise or ethical for an expert to get himself or herself into a position where the opinion can be so easily challenged?

A section of a professional society asks a person to speak at a meeting. There is no discussion of costs, expenses, honoraria, or fees. After the speaker has given the address, he or she presents the section with a bill, which includes hotel accommodations, travel expenses, meals including the dinner at the meeting, and a fee. What are the ethics of the situation?

The following example is personal experience. I had the inside of a house painted. The painter agreed that he would do the work before the floors were refinished, and then he would return and touch up any spots he missed or that had been damaged by the floor refinisher. The floors were refinished, and the painter returned and went from room to room repairing the nicks and damages and the spots that he had inadvertently missed. One room seemed to have a large number of skipped spots and the undercoat seemed to show through. The painter said he could not touch this room up. He would have to repaint the room, and he could not do that for about one week. The room was repainted without charge. The painter claimed that with certain color paints he seemed to have more trouble than with other colors, and this room had been painted with one of those troublesome colors. Were the painter's actions ethical?

President Theodore Roosevelt in 1908 said, "Every man owes a part of his time and money to the business or industry in which he is engaged. No man has a moral right to withhold his support from an organization that is striving to improve conditions within his sphere." This is a little like the old saw, "Always taking out and never putting in soon gets to the bottom."

In recent years there seems to be more attention paid to the ethics of public officials both elected and appointed. Perhaps it is clearer today than it has been for many years, that there is corruption and failure to properly protect the welfare of the nation as opposed to the welfare of a few who are close to those in power. There has always been a great deal of patronage, and those close to the officials have often seemed to obtain the most lucrative contracts. Even in industry, there are and have been favorite suppliers, some of whom have been the recipients of lucrative contracts even though they were not supplying the best and least costly product or service. On the other hand, faithful suppliers have been easier to work with at times and at other times have gone to extremes to satisfy the terms of the contract. It seems easier to go with the supplier you know rather than a stranger. Most of us have favorite stores where we believe it is advantageous to purchase rather than going to a competitor.

Political contributions were the subject of a study by the National Society of Professional Engineers (NSPE). They were concerned with the appearance that might be created if Professional Engineers were buying work by making contributions to politicians running for influential offices.

Following careful deliberation, a task force recommended, and NSPE approved, a policy. The policy states that it is unprofessional for engineers, either on their own behalf or on behalf of their firm or employers: to make political contributions either in the form of cash or services in a manner intended to influence the award and administration of contracts involving a public authority, or which may have the appearance of influencing the award and administration of contracts involving a public authority. Under the approved policy, NSPE:

- encourages, endorses, and supports the enactment of public disclosure laws which identify political contributions to federal, state, and local candidates;
- endorses the enactment of laws and rules, administered by state ethics and election commissions and professional and trade licensing boards, which are intended to assist candidates for public office and professionals in avoiding ethical and legal conflicts relating to political contributions;
- endorses the enactment of state ethics and elections commissions in those states where such commissions do not currently exist;
- advocates that state ethics and election commissions be empowered to:
    - require timely candidate reporting of political contributions made by individuals, corporations, and other parties;
    - establish campaign contributions limits based upon the nature of public office, size of jurisdiction, compensation of public official, and other factors;
    - establish a certification requirement whereby all contractors on public work certify under penalty of perjury that they have complied with all applicable campaign finance laws; and
    - establish penalties and enforcement authority for state ethics commissions in actions against parties that fail to follow the requirements of the law.
- advocate that state professional licensing boards for all professions and trades be empowered to:
    - require at the time of licensure issuance and renewal, a written certification by each licensee, under penalty or perjury, that they have complied with all applicable campaign finance laws and rules;
    - establish penalties and take appropriate enforcement action against parties which fail to follow the requirements of the laws and rules [22].

The new policy, which was approved by the NSPE Board of Directors at its January 1989 meeting, seeks to strike a balance between the critical need for public trust in the integrity of the political and public contracting process and the right and obligation of professional engineers to participate freely and actively in political campaigns. In addition the NSPE has contacted the American Institute of Certified Public Accountants, the American Bar Association, the American Medical Association, American Judiciary Association, National Association of Counties, National League of Cities, and the National Governors Association to determine whether any of these organizations share similar concerns or have made any attempts to address the issues. The NSPE has also offered to host a meeting to discuss the issues.

9
Whistleblowing

Whistleblowing is the act of going over the head of the organization to the public or to some other authority and reporting that the organization is not serving the public or its customers properly. The failure to serve may be a case of not providing sufficient safety; a case where the customer or the public is being grossly overcharged for the product or service; or a situation where waste is rampant. [The term whistleblowing may come from the sports arena where the referee blows a whistle as a signal that the game is stopped for an infraction of the rules.]

Engineers, auditors, quality control personnel, and others may blow the whistle. One of the most celebrated cases involved A. Ernest Fitzgerald, a cost analyst for the U.S. Air Force who found that there were large cost overruns in conjunction with contracts on large cargo planes. He reviewed his findings with his superiors and was instructed to forget the situation. He felt that he owed his fellow citizens a duty. Therefore Fitzgerald went public with his findings. The response of his superiors was more than displeasure with his actions. He was disciplined, suspended, and fired. This case is well documented. There were legal actions taken in his defense and ultimately he was restored to duty. His supervisors were reportedly still not happy with his activities and did everything within their power to make his job difficult to perform and as boring as possible. The Fitzgerald case surfaced in 1969 and was not settled until 1982.

The attitude of the supervisors of most whistleblowers is similar. The first task, as they see it, is not to correct the situation but to fire the messenger. In their opinion the whistleblower has been disloyal to the company. Similar situations have developed in other high government offices. When underlings have agreed to tell Congress about the true happenings in an executive branch, they have been summarily fired, and branded as disloyal. Many of these people find it extremely difficult to find new positions at equivalent levels giving rise to the view that influential people upon whom they reported have acted to blackball the individuals and to give them very unsatisfactory references. In fact some of these references suggest that those who hire whistleblowers will be viewed with great disfavor. The *American Heritage Dictionary* defines blackball as a small black ball used to cast a negative ballot, a negative vote that bars an individual from an organization, or the act of ostracizing or excluding from a social group [23]. The question of loyalty is always a difficult one. You must decide whether to be loyal to your superior, the company, the public, or yourself. When there are no problems you can be loyal to all of the entities without conflicts. When there is malfeasance either you must blow the whistle or become guilty of being a party to the act. Simply separating yourself from the activity may not provide convincing proof of innocence.

Many engineering and other societies have in their code of ethics enjoined their members to give primary importance to the public's safety and well-being. The question that most whistleblowers face is that they may have difficulty in obtaining reasonable employment after they have acted.

The *New York Times* in its October 4, 1987 issue reports the case of a whistleblower named Kenneth Prill [24]. In 1979 Prill was driving a tractor-trailer that jackknifed due to faulty brakes. He was driving for a Michigan firm. He refused to drive the rig after the accident and reported the condition to state police. The employer discharged Prill. Prill hired legal counsel and appealed to the National Labor Relations Board contending that his firing violated the Taft-Hartley Law. An administrative judge agreed, but the full board overturned that ruling in 1984, contending that the law only protected union members who acted in concert. It did not protect Prill because he had acted alone and was not a member of a union. Appeals and hearings were still in progress in October of 1987. Kenneth Prill said he would be at the hearings, and that one thing he had learned was that the wheels of justice can be very slow. Prill no longer drives a truck but is a vehicle parts salesman.

There is insufficient information here to make a real judgment call. Did Prill complain to the owners of the business and then told to drive on? Was he isolated at the scene and told the state trooper that was investigating the accident that the brakes were in bad condition? Was he aware that the brakes were in bad condition before the accident? These might be circumstances which would affect your opinion as to the merits of the case. Do you have any other questions? In your opinion what should Prill have done under the circumstances? Did he act responsibly?

Another case that was in the news recently concerned the activities of some of the engineers at the Morton Thiokol plant, where the booster rocket for the *Challenger* space shuttle had been manufactured. There was considerable concern in the minds of some of the engineers as to the wisdom of launching the *Challenger* on the cold morning in January 1987.

This concern was communicated to the management at Morton Thiokol and also to the launch command. There seemed to be more insistence on questioning the concern than in assuring that the *Challenger* was reasonably safe. The result is well known. Not only was the shuttle lost, but seven individuals perished as the result of the fire that destroyed the vehicle during its launch. The government and Morton Thiokol have since settled with the heirs of those who were lost for approximately $77 million. It should be noted that the Morton Thiokol organization was most displeased when Allan McDonald and Roger Boisjoly testified about problems with the shuttle booster. It was reported that both were assigned menial jobs. The adverse publicity caused the company to reinstate them, but Boisjoly took a leave of absence since he found the entire episode very disturbing.

It is naturally upsetting to be associated with an incident in which seven people are killed. The individuals must feel that if they had been more persuasive the launching would not have occurred and perhaps the accident would not have happened.

Charles (Chuck) Atchison according to the *New York Times* [25] was a quality control inspector at the Comanche Peak nuclear plant in Glen Rose, Texas. In 1983 he appeared before one of the regulatory commission hearings and exposed numerous safety infractions he claimed were not being corrected at the construction site by Brown & Root. They were building the plant for the Texas Utilities Electric Company.

Atchison claimed that he could not get anyone to initiate corrective action. He wound up out of work and unable to obtain a job in a similar operation. In fact he held one job for a week but was fired as a troublemaker because of his Comanche Peak reputation.

A whistleblower is apt to find little sympathy among superiors, in other companies, or in government circles. Karen Silkwood died before she could complete her efforts to vindicate herself in her dispute with Kerr-McGee. It will never be known what was true and what was allegation.

Perhaps the best-documented case of whistleblowing is the BART case. BART is the acronym for Bay Area Rapid Transit, a transportation system installed in the San Francisco Bay area. As originally conceived the system was to operate with no motorman-driver on the units. Whether this was a wise objective is not of concern. There are airport systems where several trains operate on a relatively small track and are controlled automatically. Driverless track vehicles are possible and in use. The problem relates to an alleged case of blowing the whistle on bad management, poor planning, and perhaps substandard engineering.

The BART project is the effort of three counties, San Francisco, Alameda, and Contra Costa. They banded together to create a district commission, BART, to build and manage a high-speed, rapid transit system with 38 stations and about 75 miles of track. The effort is governed by a board of directors consisting of four individuals from each county.

The work on the project was contracted to Parsons-Brinkerhof-Tudor-Bechtel. The contract for $26 million to design, install, and operationally qualify the automated train control was awarded to the Westinghouse Electric Corporation on a competitive bid in 1967.

Stephen H. Unger of the Department of Electrical Engineering and Computer Science of Columbia University and the Center for Policy Research reported on the events in the Institute of Electrical and Electronic Engineers Committee on Social Implication of Technology newsletter (No. 4, 1973) [26]. Three engineers on the project, Holger Hjortsvang, of the maintenance section, Max Blankenzee, a programmer analyst, and Robert Bruder, an electrical engineer, monitoring various phases of the project, had each been critical of the management and the handling of the project and had complained to their superiors. None of the three had obtained the corrective action they thought was necessary. They then approached Daniel Helix, a member of the board of directors, and not only complained to him but also presented him with data.

Helix informed the other members of the board and top management and apparently notified the press that there was some controversy. Subsequently, late in February 1978, the board held a hearing where the contract management was allowed to defend its action and present its data. The board voted 10-2 to uphold the opinion of the contract management.

Up until this time the identity of the three engineers had not been made public. This was done early in March and the three were given the choice of resigning or being fired. They refused to resign and were dismissed with no written reason being given.

The California Society of Professional Engineers, their transportation committee, and at least one of their chapters investigated the situation. They concluded that the three engineers had acted in the best interest of public welfare in disclosing to the BART board of directors problems associated with train control, systems management, and contractual procedures, and that there were other instances of poor employment practices at BART. The matter came before the state senate following more investigations, all of which confirmed the claims of the three engineers.

Blankenzee charged that BART officials intervened several times to prevent his being hired by other organizations on the basis that he was a troublemaker. The engineers were having great difficulty in obtaining employment at responsible tasks.

The IEEE interceded as a friend of the court in a suit filed by the engineers seeking compensatory payments from BART. The gist of the IEEE submittal was similar to the ethical code in that it was wrong and improper to punish engineers, dismiss them, and deny them reasonable employment because they had properly discharged their duty of protecting public safety and welfare. Eventually a settlement was reached and the engineers received some compensation.

One interesting sidelight to this affair was that the press had tried to reach one of the engineers before the board hearings and had been rebuffed. In addition the engineers had gone over their supervisors not to the outside, but to the board of directors. What was most interesting in this entire case was not only that the engineers were vindicated by further analysis and audits, but the first runs of the transit system were unsuccessful in properly controlling the trains, and drivers had to be used for safe control.

Several questions arise from these cases. Were those classed as whistleblowers properly performing their jobs? Were they discharged and harassed for whistleblowing or were they troublemakers who had no justification for the claims they made? The BART case was most unusual because of the entrance of the IEEE and the California Society of Professional Engineers who maintained that the individuals were properly performing their jobs and protecting the public. The IEEE has two bylaws endorsing and encouraging adherence to professional standards.

Bylaw 112.1

> A member of the IEEE may be expelled, suspended or censured for cause. Cause shall mean conduct which is determined to constitute a material violation of the Constitution, Bylaws, or Code of Ethics of the IEEE, or other conduct which is seriously prejudicial to the IEEE.

Bylaw 112.4

> The IEEE may offer support to any member involved in a matter of ethical principle which stems in whole or in part from such member's livelihood, and which can jeopardize that member's livelihood, compromise the discharge of such member's professional responsibilities, or which can be detrimental to the interests of the IEEE or of the engineering profession [27].

All those who must make decisions must ask, when does the situation become bad enough to make it necessary to become a whistleblower? The most notorious individual who has been a whistleblower of sorts is Ralph Nader. In the interest of public safety, health, and well-being, he has done his utmost to expose situations in which some manufacturer or seller has not been serving the public safety and interest. Nader is not a whistleblower in the true sense of the term since he is not reporting on work done in a company in which he has been actively employed. He is an outsider and targets various products and companies. Another whistleblower of sorts is the Consumers Union which publishes *Consumers Reports*.

The ethics of operating a cooperative which tests products and reports to its members the relative value of these products, is not an open-and-shut case. The difficulty with the Consumers Union approach is that the organization would be open to charges of being unfair and dishonest were their reports not based on factual evidence. This places a heavy burden on the operation.

Another area of whistleblowing in government is the U.S. Consumer Products Safety Commission (CPSC). This organization has been a major thorn in the side of many manufacturing and consumer product sales organizations. That it has done some good is unquestionable. Some administrations have questioned whether this type of organization should be sponsored by the government since it has been charged as being antibusiness. Whether government should favor business over the consumer or the consumer over business is open to question. The government should, many believe, favor safety. It can also be claimed that working to establish standards of safety and minimum performance favors one group against another. Hence the feeling, on the part of some, is that the standards-sponsoring activity is misplaced when it falls into government hands. Do you believe that it is proper for the government to establish standards relating to safety?

When the government established a tire safety rule and made it unlawful to use bald tires or tires with little tread on the highway, there were some who objected. The government has no right to prevent them from using tires longer. Furthermore, some stated, nobody had the right to prevent them from risking their lives, if they so chose. It was pointed that although they might have the right to risk their lives, they had no right to endanger oncoming motorists or pedestrians. Do you have other opinions regarding the government's right to enforce safety standards or standards of economy in fuel usage?

Some of the most widely adapted standards in the country concern safety. These include standards for drugs, foods, automobiles, building codes, electrical installations in buildings and homes, and safety standard for electrical equipment, furnaces, and heaters. These are so widely adapted into law and so widely accepted that there is little discussion of them. The concept that the early rotary blade lawn mower was unsafe was challenged by the industry. It was only after a long time that the industry took up the task of establishing a standard and making safer mowers. No device of this nature, no power saw, no knife, and no automobile can be made that is absolutely safe. There are only ways of making the devices safer.

Despite the statistics that wearing a seat belt properly provides greater safety to vehicle occupants, there are many drivers and riders who do not follow the practice of buckling up. In the event of an accident, should the individual who refused to buckle up be protected by insurance to the same extent as an individual who wears seat belts?

Gil Courtemanche in the *Internal Auditor* of February 1988 discusses the ethics of whistleblowing from the viewpoint of the internal auditor [28]. He remarks that a third of those he questioned on the ethics of an internal auditor blowing the whistle on a corporation thought that this should not be done; another third said that they would blow the whistle; and the third portion said they felt they should, but were unwilling to bear the consequences. Courtemanche argues that there are many reasons for an auditor not blowing the whistle, and that auditors are fully in compliance with their duties, if having discovered a wrongful action, it is reported to the supervising auditors or other superiors. He justifies this argument by relying heavily on the fact that internal auditors have a foremost responsibility to their clients.

Courtemanche is firm in his belief that audit files should be so managed and so documented that they do not have any possibility of being called into court and convicting the corporation of malpractice.

He points out that when a corporation finds an employee guilty of some illegal practice, stealing, or other indictable crimes, they usually do not prosecute but instead terminate the employee. He carefully skirts around the situation where the corporation is under the guidance of an officer who is actively engaged in an illegal action and advises the internal auditor to consult with the company attorney when in a difficult situation. The ethics of having knowledge about a crime and not reporting it to the authorities is not a breach of ethical behavior according to Courtemanche.

The article highlights the fact that internal auditors are not compelled to disclose to outside auditors that there are irregularities or illegalities. Courtemanche further indicates that the outside auditors, usually CPAs, have a weapon which they can readily use. The weapon is refusal to provide a clear endorsement of the annual or quarterly statement.

One of Courtemanche's reasons for recommending against blowing the whistle is affirmed by the National Commission of Fraudulent Financial Reporting. The Treadway Commission of October 1987 stated that, "for the most part, those making disclosures of alleged wrongdoing or illegality find little or no protection. Whistleblowers almost inevitably pay a heavy price. With few exceptions, they are driven out of not only their jobs, but also their professions"[29].

One argument is that the price paid for whistleblowing is too high and another is the question of loyalty to clients or employers. If these conclusions are accepted, there is a greater loyalty to others than to honesty, truth, the public, and oneself.

This same attitude is seen in several cases where there have been questions of integrity in government operations. Was the loyalty of government employees to their superiors greater than their loyalty to truth and to the country? If they lied to protect their superior, they were called loyal. If they told the public the truth which was at odds with their superiors' interest, they were branded turncoats. Never mind the fact that the question was never asked which loyalty takes priority. In fact, the unapproved release of knowledge of unlawful conduct or employee fraud to outside auditors by inside auditors is interpreted by Courtemanche as a violation of Articles II and V of the Code of Ethics of the Society of Internal Auditors Incorporation.

As opposed to the codes of some of the major engineering societies, Courtemanche seems to say that the auditor's code of ethics requires the internal auditor to have little responsibility to the public. The external auditor, as previously noted, has the privilege of withholding a clear endorsement of the annual or quarterly statement. It this sufficient to satisfy your requirements under all conditions?

The Code of Ethics of the Institute of Internal Auditors states that the members shall "conduct themselves so that their good faith and integrity should not be open to question . . . and they will promote the highest possible internal auditing standards to the end of advancing the interest of the company or organization." The code is shown in Figure 6.3.

In reviewing this code it is most noticeable that it is missing some items which are presented in other codes of ethics, primarily a responsibility to public welfare and safety. The engineers' codes hold public safety and welfare to be a primary responsibility.

The Code of Ethics of the Institute of Internal Auditors is similar in some respects to that of the American Bar Association in that members are not to do anything prejudicial to the interest of their clients. The codes are different in that the members of the bar are not restricted from activities, not related to the case in which they are representing the client, which are detrimental to the interests of the client.

Under this interpretation lawyers can decide that in representing their clients, they can interpret the law to exonerate the client, but at the same time they can work to have the law amended. This way, clients may no longer be able to practice the same behavior without running into a conflict with the law.

There is also no need for lawyers to disclose information about clients to officials. In this way both the internal auditor and the lawyer may be in possession of information which is prejudicial to the public interest, and neither is asked to disclose the information, according to their code of ethics.

Lawyers are to refrain from disclosing the secrets of their clients. Notwithstanding this there have been cases of large law firms working with corporations to prevent them from being taken over by unfriendly suitors, yet turning around and helping a third party take the firm over. These actions were recently discussed, and the final decision was that the law firm had acted in a proper and ethical manner.

Earlier it was asked whether it was proper for engineers leaving one company to disclose to their new employers material known about the former. If the disclosure related to patent efforts, it might result in patent litigation. If the disclosure related to plans of the old company to bring out a new product not protected by patent, the new employer would be able to introduce a similar product at about the same time. If it related to a bid strategy and material that was about to be introduced into a large government bid, this might place the new employer in a very advantageous position. Is this proper and ethical?

While I was engaged in the hearing aid business the transistor was invented. It was obvious to many that the transistor would make it possible to reduce the size of the hearing aid and reduce its need for as much power as was used by the current electron tubes. Ultimately this would make it economically possible to use a smaller battery and reduce the size of the aid. Another advantage that transistors promised, and subsequently proved, was that the transistor was more reliable.

Sonotone Corporation embarked on a program of investigating the application of transistors to hearing aids. All of the other companies in the hearing aid industry were also interested. They were encouraged by the Raytheon Corporation which supplied all the other companies with vacuum tubes and which promised to supply them with transistors as soon as it was able. Sonotone made all of the tubes it needed and supplied no one else. Raytheon let the other hearing aid manufacturers know that Sonotone, the largest in the business at the time, would not get any transistors from Raytheon. It also let the other manufacturers know that they had no requests for information from Sonotone, and that since Sonotone had just brought out two new models, it was obvious that they had no intention of going into transistors.

Transistors, although they were advantageous, had one disadvantage at that time; when used in the gain stages of the hearing aid amplifier they were noisy. In fact, so bothersome was this noise level, that the introduction and distribution of transistors by Raytheon had to be delayed. Sonotone knew this. They had found a source for a power output transistor, which in combination with tubes in the gain stages of the hearing aid, resulted in an aid with much better life and much reduced battery demand. There were a limited number of battery suppliers to the industry. These people knew that there were no plans for a change in battery complement at Sonotone, and this information also seemed to be known in the industry. The Sonotone plan to use a power amplifier transistor required only one simple change in the chassis of the hearing aid and did not materially reduce its size or weight.

The operation was known to only a half dozen people in the corporation, and the change in the instrument was called the introduction of a new switch.

The introduction of the first hearing aid with transistors by Sonotone was heralded and obtained much publicity. The major reason was that this was the first application of a transistor to a consumer product; in fact the first application of a transistor to any consumer, industrial, or commercial product. The publicity was excellent, and the sales were good. The remainder of the industry was able to follow a few months later with material from Raytheon.

A few weeks before the new switch was installed one of the technicians in the production plant notified us that he was leaving to go to a competitor. We were quite certain that he was not privy to the information on the progress of a change-over to the transistor model so we ushered him out of the plant with severance pay. I met him about a year later and was surprised to learn that he had no idea of why we let him go immediately, rather than letting him go at the end of his two-weeks notice. I did not explain.

What we were worried about was whether he might discover, in the course of his last two weeks, the transistor secret. If he had known, what should he have done on an ethical basis insofar as letting his new employer know of the impending introduction of the transistor hearing aid?

10
Loyalty, Ethics, and Job Hopping

There are many kinds of loyalty including loyalty to bosses, customers, suppliers, companies, fellow employees, country, and self.

Some codes of ethics are specific about the responsibilities that engineers owe to a number of entities and to themselves. There are the responsibilities to the public and to keep abreast of the technology. These are intertwined since engineers cannot do a good job unless they keep abreast of the progress in the field.

The American Society of Civil Engineers (ASCE) Code of Ethics states that engineers shall hold paramount the safety, health, and welfare of the public in the performance of their professional duties.

This code also states that engineers shall continue their professional development throughout their careers, and provide opportunities for the professional development of those engineers under their supervision.

A similar portion of the Code of Ethics of the Institute of Electrical and Electronic Engineers (IEEE) states that members shall maintain their professional skills at the level of the state of the art and recognize the importance of current events in their work. This code also states that members shall assist their coworkers in their professional development.

It may be hard to equate loyalty to company and to fellow employees when engineers hop from job to job. Is it real loyalty when you take a job with a competitor? Is taking a job with a competitor acting as a faithful agent of one employer? If you take a job with a competitor how much of the knowledge of the job and the company practice do you have the right to take with you? The new employer is expecting to get a well-trained engineer or quality-control specialist with knowledge of the field. Does he or she have a right to all the knowledge that you bring over from the other job?

In another instance examine the engineer in private practice who works as a consultant. She accepts jobs from a number of companies. She has been working for several years with Company A that makes a widget. Company B also makes widgets. Company B asks this engineer to help them. Can the engineer contract to help Company B while working for Company A?

The question also arises when a consultant works in an industry and develops the knowledge that permits him to solve problems for members of the industry. A company for whom he has not worked wants to use his services to solve a problem. They ask him what arrangements they can make. Can they employ him for an hour or two and expect to get the fruit of many years of effort? What is a fair price for his services? Should there be a minimum fee for services regardless of the length of time used? All that Company B would like to acquire is a solution to a problem which this engineer has seen solved more than once. Should there be a time or a contract price for the job? The engineer can solve the problem because he has solved it many times and knows exactly what must be done. Company B on the other hand might have engineers work on the problem for several years and not solve the problem.

A similar instance is often quoted. Several riggers were asked to move a piece of equipment into a pit about 15 inches deep. The prices quoted by several of the riggers included the removal of the roof and the cost of bringing in several large cranes. One rigger said that he would do the job without bringing in the cranes and without raising the roof. His price was very low. He was asked for his technique and replied, "Give me the job at my price and then you will learn the methodology." His technique consisted of filling the hole with blocks of ice and moving the equipment onto the top of the ice. As the ice melted, the equipment slowly sank into the hole. All that was necessary was a pump to remove the water and a steam line to accelerate the melting where it seemed slowest, so that the equipment would not tip. Here was a rigger who knew how to do the job most expeditiously and economically.

A similar situation occurred during my employment on a contract to design and deliver power supplies for missile sites. The contract contained a clause that required that the device pass a blast shock test on a specific machine before it could be approved for construction. The contract further stated that the contractor must submit proof that the device would survive the blast shock test before the test was performed. There was only one blast shock test machine available.

When the power supply was almost ready, the clause requiring the prior proof was noted. It had been overlooked. Several engineering firms were called in to quote on the task of providing proof before test. We were shocked at the prices. This was in the 1960s. The prices ran between $25,000 and $50,000. There was no money in the contract for such an expenditure. A simple inexpensive procedure was suggested. The answer by each and every engineer and the contract administrator was that the prime contractor would never buy such a simple test. No one could find any fault with the suggestion so we proceeded. The test proved that the unit would not have passed the blast shock test, but it also showed where the weaknesses were. The mechanical structure was strengthened and the test now indicated that the structure would pass a blast shock test of twice the magnitude. The entire test and the fix cost less than $1,000. The program as performed was submitted and the proof was accepted. When the blast shock test was performed there were two units to be tested, ours and a competitor's. Our unit was tested first and, as we had expected, it passed. The competitor's unit failed. When the crane was used to remove the competitor's unit, someone neglected to remove all the hold-down bolts and the blast shock table was severely damaged. It was out of service for a year, and the competitor could not deliver any units for that period.

Suppose some organization has a similar problem and came to me and said that they wanted a method of proving that their product would survive a blast shock test. What would be a reasonable fee? Actually the entire procedure can be explained in minutes. Is it ethical to ask me to give this information away for $10? Is it ethical for me to charge for a day's work to explain the procedure when it can be explained in minutes? When this problem occurred the cost of solving it was $1,000 and the lowest quotation had been $25,000. Even though one can calculate much more rapidly today, the solution is still worth quite a bit to anyone needing to prove the integrity and strength of a unit.

A similar situation occurred when a college professor was asked to consult on a product which required some special materials. This professor might have been able to explain in an hour or two, the actual procedure that the company should use, the materials, and how to mix and treat them. The company asked him to come to the plant for a half day and give them aid. He said that he felt they were asking him to give them his years of experience for a paltry half day's work. The hourly fee was small. Was it ethical on the part of the company? Was it ethical on the part of the consultant?

These problems are discussed under loyalty, because there appears to be a conflict between loyalty to self and loyalty to a client. Where does the one conflict with the other? How would you suggest that these situations be resolved?

One possible solution is a contractual agreement based on both sides knowing the actual situation. The consultant and the client agree to a certain amount of time and a certain fee. What happens when the client gets all his or her answers shortly after the contract is initiated and cancels all further work thereby leaving the consultant with a small fee? There may be a question as to whether this is ethical. Can you suggest a fair and equitable solution that treats both sides fairly?

Being ethical includes being loyal to yourself. It includes advancing your education in accordance with the requirements of the codes of ethics. This may be accomplished by taking or teaching courses, writing papers for publication, developing new methodologies, making inventions and discoveries through analysis and research, and/or studying the latest papers and publications.

All of these activities require that the engineer keep abreast of certain facets of the state of the art and thereby learn, and in some cases, contribute to the development of the technology.

Some simple methods of performing tasks were mentioned. Sometimes the release of information that something has been accomplished is sufficient to allow some of those who are familiar with the field to duplicate the event without reading the details of what the original inventor had done.

In 1945 the United States dropped nuclear bombs on Hiroshima and Nagasaki. In a matter of days those familiar with nuclear technology recognized that a breakthrough had occurred. There had been a reduction to practice of the theory that maintained that there were methods to release nuclear energy. Those in the trade rapidly surmised what had been done, although they could not be certain of the details of the methods employed. The Smith Report gave further details. Scientists and engineers in Nazi Germany had been working on the same problem, but had not had the concentration of effort and funds that permitted the United States to accomplish the feat.

Most of the details were surrounded by secrecy and only a limited number of individuals were allowed to know the true details of this device. The Hiroshima and Nagasaki bombs released large amounts of radiation which were detrimental to many people. Earlier explosions of smaller pilot bombs in New Mexico exposed animals and people to injurious radiation. These aspects of the bomb were not thoroughly appreciated at the time, and greater exposure was allowed than should have been.

The amount of secrecy that surrounded anything associated with this atomic bomb effort can be illustrated by the following event which I witnessed while in the lobby of an agency engaged in classified work. A courier came in and asked the receptionist for Dr. K. Shortly thereafter while he was waiting in the lobby with his briefcase handcuffed to his wrist, Dr. K.'s secretary came to the lobby and asked the courier what he wanted.

He said he had Dr. K's paper for the AXZ symposium in his briefcase and that it had been reviewed by the censor.

The secretary asked for the paper and stated that she would take it to Dr. K., who wished to review it to see what had been deleted before he went to the seminar.

The courier said he could not give it to the secretary.

The secretary said she had clearance and was entitled to handle the paper and in fact she was the one who had typed it.

The courier's answer was, "I have been given my instructions. I am to accompany Dr. K. to the symposium. When it is time for the presentation I will remove the paper from my briefcase and place it on the lectern. Dr. K. will read the paper that I place on the lectern. I shall then remove it from the lectern, lock it in my briefcase and return to the security office where I will return the paper. It will not be out of my control till I return it to the security office."

The courier knew his responsibilities and loyalties. There was only one type of conduct.

With company-classified information which comes into the hands of many engineers, it is difficult to abstain from carrying this information from one job and from one employer's domain to that of another. To some extent the same must be true of material classified by the government. How can classified information and the knowledge that comes from being exposed to it be prevented from traveling with the engineer from job to job?

When people move from one job to another, new employers expect to obtain individuals with sufficient skills to do the work they will be assigned. In the case of engineers or others with specific technical experience, new employers will expect that the individuals will have acquired skill and experience while on their previous jobs. New employers do not expect that the engineers are incapable of doing their job nor that they had failed on the previous one.

There is always some question of loyalty when changing jobs. The individual who is moving is leaving a company for whom he had been doing beneficial work. He is leaving the company and the other employees to take a job with another company which, if it is in the same product line, will be taking work away from the company he is leaving. He is therefore really changing sides in a game of economic war. He must, by necessity, be taking along trade secrets and an intimate knowledge of what is going on in Company A and can not help but carry some of this knowledge to Company B.

If, on the other hand, the engineer is leaving a company that is engaged in designing, manufacturing, and selling loud speakers for high fidelity equipment and is joining a company that is engaged in designing, manufacturing, and selling thermostats, she is carrying over to the new company skill and not intimate knowledge of a product in direct competition with her former firm's endeavors. Is Employer B likely to benefit from the engineering skill that was developed by the engineer while working at Company A? How much of the art and progress at

Company A can the engineer ethically divulge to B? How long a time should pass before Company B can discuss with the engineer some of the happenings and the progress at Company A? Is this a period of days, weeks, months, or years?

Assume that the engineer has not been engaged in design, but in the operation of a quality control operation. Could Employer B benefit from the quality control and manufacturing engineering skills developed at Company A and by the systems developed by the engineer while at Company A? How much of these skills can the engineer ethically divulge to Company B?

In the event there had been a layoff at Company A or a situation had developed which resulted in the termination of the employment of the engineer, does the question of loyalty still apply and do the restrictions of transfer of information change?

There is another situation relating to loyalty. An ethical person can seldom serve more than one master at a time. This brings up the question of accepting gifts from vendors without the full consent of the senior members of the corporation. Even with their consent there may be situations where it is not only unethical, but it might be unlawful.

Many companies prohibit their employees from accepting gifts from vendors. They notify vendors that this is the company policy. For example, the question may arise in the course of the procurement operations that Jones, the procurement manager, and Burke, the engineer for Apar Manufacturing Company must visit a plant to assure themselves and the company that Vendor Company is capable of performing under a proposed contract. There is a high probability that Vendor product will be conforming and delivered on time. The trip and the purpose are known to both managements.

Vendor wants this business and proposes that it will do all within its power to facilitate the visit and the survey. Vendor suggests that it will send its corporate jet to a convenient airport, transport Jones and Burke, supply a car, put them up in a nearby motel where it maintains a suite, and supply meals and entertainment during the period that Burke and Jones are at their site performing the survey. This is a major contract and the survey may take two or three days. Is there any impropriety? Is there any ethical reason not to let Vendor do this? Are there any limits that you would place on this offer of hospitality?

Apar Manufacturing has a policy of reimbursing all personnel for travel at a standard rate per diem. How should this be handled? Who should be advised of this offer and its acceptance? Do you have any reactions to a situation such as this? If so, what are they?

I worked for one company where engineers and others patronized the company cafeteria for coffee breaks. Engineers John and Sam were discussing ways of supplementing their income. One way was to do consulting work for other companies. The corporation was involved in the design, manufacture, and sale of electronic, electromechanical, and communication equipment in the military and civilian markets. Both agreed that the broad spectrum of activities related to the company's work made it almost impossible for the engineers to accept any consulting work with organizations involved in the electronic industries. They could teach, consult for a construction corporation, an electrical machine or appliance maker, work in a field unrelated to the company's business or work for a company making a product that we were unlikely to buy or make.

The ASQC Code of Ethics in Section 2.5 specifically says that the ethical person "will not accept compensation from more than one party for the same service without the consent of all parties. If employed he will engage in supplementary employment or consulting practice only with the consent of his employer." This indicates that such activities would be permissible only with the consent of the corporation.

John and Sam's conversation then switched to the question of courtesy and the proper practice regarding visits to vendors and from vendors.

If a vendor visits the company offices and meets with an employee, it seems appropriate during lunchtime, to have lunch with the vendor. What is the proper practice?

If it is decided to eat at a local restaurant should either party pay the entire check or should the employee let the vendor pay his or her own way? What if they decide to eat in the company cafeteria?

In some companies employees are encouraged to take visitors to lunch in the company cafeteria. They pay their own tab but sign a chit for the guests.

Is it proper to accept an invitation to a vendor's lavish Christmas party? Would it be proper for employees and their spouses to accept similar invitations?

Is it proper for employees to accept vendor's invitations to dine at an ordinary or an expensive restaurant? Are there conditions under which it might be proper and others under which it would be improper? If the employees' spouses were included in the invitations, would the propriety of the situation be changed? Would it be proper to accept if the employees and their spouses were invited to join vendors and their spouses at similar occasions?

Would it be proper for employees and their spouses to spend a weekend at a university football game, all expenses paid, including food, lodging, tickets, and transportation provided by the vendor? Does it change the situation if the employees' spouses are not included? Is the situation any different if World Series tickets are involved rather than football tickets and there is no travel?

Is it proper for employees with or without their spouses to spend a week or days at a resort or on a cruise with all expenses paid by vendors? Does it make any difference if the upper management is aware of this arrangement?

There was another person at the table and John and Sam suddenly noted how quiet and cold the third individual became. This individual and his wife had been on a cruise with another corporate officer and his wife. A contract had recently been signed with a large insurance company covering much of the corporate insurance. John and Sam thought it unlikely that the two couples would travel together. Had the insurance company provided the cruise? They had no proof and never investigated but strongly suspected that there had been a payoff. Perhaps this was the best deal anyone could make with an insurance company, but even if this were so, were the two men ethical in their acceptance of the gratuity, assuming that this had been a gratuity? Is there a limit which transcends the bounds of ethics?

Would different decisions be made if the individual was a design engineer, a procurement person, a quality-assurance engineer, or a quality-control manager? Does the sex of the individual have any effect on your answer? Would the answers be different if the individual representing the purchasing organization was a vice president, president, or other executive of the corporation?

In another situation which I heard about, Frank's daughter was suffering from a serious but not well understood disease. The physician with the most knowledge of this disease was located at a considerable distance from the area where Frank lived. A vendor offered to have a private plane fly Frank, his daughter, and his wife to the area for a complete examination. The expenses of the trip would be borne by the vendor. The physician's fees were part of a medical payment plan and therefore did not enter into the discussion.

Was this an ethical offer? Would it be proper for Frank to accept this offer? As the employer, what would be your opinion? As a competitor of the vendor, would you have other opinions?

There are organizations that have rules concerning gifts and courtesies. One rule is that if you consume it you can accept it. This is designed to apply to a meal and not to a case of scotch. I have heard of procurement departments that at Christmas time called certain vendors and advised them of the home addresses of procurement personnel and their favorite brands of liquor. Would you class this as strictly within the ethical constrictions of your company's rules?

Consider the range of gifts and courtesies. These include inexpensive and expensive calendars, magazine subscriptions, luncheons, dinners, cruises, and resort vacations. There are even more expensive gifts. It is difficult to decide where the limit ought to be, and the result can well be that the rule is left open and unanswered or set so that no amenities may be accepted.

Even when the rules are set, there are great difficulties in enforcing them with an even hand. Can you refuse a ride in someone's car? Can you refuse a small snack or coffee in someone's office? When these situations arise it is often impossible for the individual to ask for guidance. Engineers and company representatives must make up their minds on how to handle the situation on the spot. Should they report the courtesy to their superior? If they feel that there has been some attempt to compromise them and the company, the answer to this question must be in the affirmative. Should the incident be reported?

Sometime back I was employed by a small company listed on the American Stock Exchange. It was small in the sense that the entire home office engineering and manufacturing operations were housed in one building and the president of the company knew each of the engineers on a first-name basis. The president was an imposing individual, a super salesman of whom it was said that he could sell iceboxes to eskimos.

My job was that of a supervising engineer reporting to an assistant director. He reported to the departmental director who in turn reported to the president. The director was not a vice president of the company.

The director was, to my way of thinking, totally incompetent and concerned himself with trivia, to the extent he would carefully read every document on and in the desk of an absent engineer and answer any letter that was more than one week old. His answers would include the sentence, "I found this letter on Harry Pines' desk and since it had not been answered, I am sending you this reply."

The director carefully issued his handwritten orders to engineers on preprinted memo forms. These included orders to start, curtail, accelerate, or stop work on specific items. Although he appeared organized he was, in my opinion, totally devoid of knowledge regarding our product and the company operations. In addition he was one of those individuals who was of the opinion that his way was the one and only way to operate.

There was general dissatisfaction among the engineers who communicated their unhappiness to a vice president of another division. He pointed out that Dr. Portland did not report to him but to the president, and that if corrective action was to occur someone would have to convince the president that there was a serious problem.

The two assistant directors although unhappy were unwilling to sit down with the president, perhaps feeling that they would be unable to make a plausible case.

I had considerable contact with Dr. Portland and was asked if I would meet with the president and present our case. I agreed because, not only was I unhappy, but I also had an offer of another job and felt that I had to make up my mind to stay or go in the very near future. In addition I had more than 20 "do-it-now" orders from Dr. Portland which I had carefully ignored while documenting the effective actions I had taken to initiate corrective action. I had also been witness to the day-by-day activities of Portland.

An appointment was set up for me to meet with the president. The vice president arranged the meeting and advised the president of the general dissatisfaction that I would be discussing. The meeting was scheduled from 2:00 to 2:30. At 2:00 I was admitted to the president's office. I explained that I had another job offer and had sought it out because of my unhappiness with the present administrative setup in our department. I explained that I felt that it would be impossible to continue very long under the administration of Dr. Portland, but I wanted to talk to the president because I felt that I owed some loyalty to him, to the company, and to my fellow employees. I had been there some seven years. If I were to accept the other job and say nothing it would not be fair to these people and the company. I had therefore asked to speak with him and explain my opinion and what I had witnessed. I was fully aware of the fact that the president had searched for and selected Portland.

We reviewed my complaints point by point and at the same time I presented the evidence. Although he did not indicate any agreement, he made no effort to disagree with my presentation. The handwritten orders were so patently wrong and my corrective actions were so appropriate that there was little room for comment. The president was more familiar with our product and our customers than Portland. He was a salesman but the product was an important part of his knowledge.

The desk searches disturbed him and he wanted to know how I was aware of them. I explained that during the workday Portland would sit down at the desk of an absent engineer and methodically go through the papers on top and in the drawers. These were not searches for specific items. Portland would read all papers.

The president was interrupted with several phone calls and also called his secretary to postpone a 2:30 meeting, then a 3:00 meeting and so on. Toward 5:00 the president had his secretary bring in the letters of recommendation he had received before he had hired Dr. Portland. There were quite a few. He started to read them to me. After the first few he said, "They do not sound the same as they did before I hired Dr. Portland." Each one was similar to the others in saying that Portland had worked in a specific job at their company, but it had not been research and development. He should make a good director of research and development. In addition he had been at each organization for only a few months.

The meeting ended about 5:30. My half hour presentation had stretched to three and a half hours. For the last hour I had a distinct feeling that the president, the super salesman, had been trying to justify his actions and fight his way out of a corner. I was convinced that he knew he had made a mistake. He asked me to delay my decision to leave. He made no commitments to me.

He did ask me how I would feel if Dr. Kantor were to replace Dr. Portland. I replied that we had discussed this among ourselves and had decided whose car we would use to go to a government facility about 100 miles distant and apply for jobs there. How about Dr. Green? I replied that he was well respected, and that as far as I knew, all of us would be pleased to work under his direction.

When I returned to my desk at 5:30 the engineers and the vice president who had arranged the meeting were waiting for me. They asked if I still worked for the company and what had happened at the meeting and why the meeting had lasted so long.

Among the questions you might ask are: Were my actions ethical? Did I do the correct thing? Was I being loyal to my boss? Was I being loyal to the company? Should I have kept silent and just left? Was there some other action I should have taken? Explain your decisions by reference to one of the engineering society's code of ethics.

Within 10 days Dr. Portland left. Dr. Green was appointed his successor.

11

Design Ethics and Codes of Practice

Engineers in different activities face different problems during the course of their professional careers. It is obvious that the procurement people face the most serious temptations. Engineers in different specialties must face different sets of problems, if the experience of the professional societies is an indication. As noted elsewhere, the Institute of Electrical and Electronic Engineers had no reported incidents of violations of the code of ethics, while the American Society of Civil Engineers reported a substantial number had occurred each year. The American Society of Mechanical Engineers (ASME) was not willing to release any information. They have had one very serious case resulting in substantial awards to the plaintiff. The case involved an interpretation which stated that the Hydrolevel system of stack vent closure did not comply with the requirements of a standard. As a result the Hydrolevel system was not profitable. This case resulted in a substantial loss for the ASME. The American National Standards Institute settled with the plaintiff, Hydrolevel, for about $50,000 long before the case went to trial. The ASME suffered a loss of millions. (See additional references to this case in Chapter 3.)

The ASQC has had several incidents referred to the Ethics Committee, but these did not involve the society in a legal suit.

It would appear that engineers involved with product design and the sale of engineering services, particularly those who are in the consulting or private practice field, are the most likely to be tempted and involved in matters challenging their ethical conduct. The ASME case involved standards and restraint of trade by interpretation of standards. In the Hydrolevel case, those involved interpreted a standard in a manner that seriously and adversely affected a competitor.

Design engineers become involved in standards since their design must comply with the standards that are in effect in the marketplace. They also become involved in patents, and must be aware of the state of the patent art, so that they can take advantage of their company's position and avoid infringement of patents belonging to others. There is also the possibility that design engineers will become involved in patent litigation, either in helping their company defend against a charge of infringement or in pursuing an infringement of the company's patent(s) by a competitor.

The knowledge of the state of the art in patent litigation is something design engineers carry with them when they leave one company and join another. This can result in being involved in a suit between the two companies. In one case the designer of the Marchand calculating machine left that company and started the Friden Company with a competing machine. The Marchand used a forward and reverse rotation to perform the calculations. The Friden was unidirectional and operated on a new set of principles and patents. Every series of events does not have so happy an ending. Design engineers might invent a new method of designing a product or a new process for manufacturing which is not patented. The new employer expects to have them design similar machines. Is there any limitation on what design engineers should disclose about their last employment?

In the previous chapter, we discussed a phase of a code of ethics which suggests that employees or engineers shall not be paid by two organizations for the same work. What is your opinion of the proper procedure for design engineers in private practice to follow in their billing for several jobs under the following conditions?

An engineer is involved in designing a bridge for the State of New York. He is asked to design a similar structure which is to be located in Pennsylvania. Is there any conflict of interest? If he bids on both jobs and obtains them at a fixed fee he stands to obtain a substantial profit because some of the work will be applicable to both jobs. This may allow him to bid the second job at a lower price. Is there anything unethical about this? It is also possible that both jobs are being bid at the same time and the engineer is not certain of getting either. If he is to charge for each job on the basis of time spent, how should he apportion the charges? Is it permissible to charge both states for the time it would take to design each bridge if the other was not being designed?

Suppose the engineer develops a procedure for doing the structural calculations. The design work would normally have taken four more weeks using the commonly known method. The new method took three weeks to develop. In charging the two states the engineer has several choices: charge each state for the four weeks normal time; charge each state for the three weeks of design time; charge each state for one and a half weeks splitting the time for the procedure development.

What should the engineer charge to future contracts when making his bids? Is there some possibility that the engineer is violating a rule of his society's code of ethics by charging two or more employers for the same work?

The same question can arise when an engineer performs a contractual task for an organization and is charging for time spent. This engineer has done quite similar work for one or more organizations in the past. The work is recorded and took 75 hours to do. She is performing a similar task for a new organization. Should she reproduce the work and give it to the new organization without charge? Should she make a nominal charge? How should she bill for this report which contains work which took 75 hours and was done while working on a fee-for-service time for a prior customer organization?

The same matter may come up when an engineer works in conjunction with a lawyer to represent a plaintiff or defendant. The subject of one product liability trial was a product that had been the subject of a previous trial. It had also been manufactured by the same company and had failed in the same manner causing injury to both plaintiffs. The first time this item came up it took several days to develop the theory and presentation. The original plaintiff paid for the work. The new client needed an almost identical report. The development of the new report will take hours instead of days. Suppose it takes two hours. Is it proper for the engineer to bill two hours, four hours, six hours, eight hours or two days?

I have a letter in my file which I have saved as an example of a practice which I consider a case of bad manners. It has other implications. The letter is from a plaintiff's attorney to whom I rendered a report relating to a serious accident. The letter in brief states: "The trial of the *John Doe* v. *Richard Roe Corp.* case is scheduled to be held in the Supreme Court of the state of . . . on the 17th of March next. Please be prepared to appear and testify at the court on that date or on a date to which the case is adjourned." I have never heard another word from the lawyer involved. The case has been settled. Do you have an opinion as to whether this was a case of ethical failure or just bad manners on the part of the lawyer? You will note that the letter stated that the recipient was to be prepared to appear and testify. Was that a work order? Should the writer have billed the law firm for preparation time? Would that action have been legitimate?

The situation just noted was a case in which the engineer was engaged in private practice. It did not present a situation in which there was an employer and a consulting agreement with another firm.

The same question might arise in respect to travel time and expenses. In many instances the engineer must travel to a distant location to survey a site or view a piece of equipment. The travel time and the cost of travel is billable to the organization for which the work is being done. Suppose there is a trip from an East Coast office to a site in Los Angeles and also another in San Francisco and a third in Oakland. The engineer arranges the trip so that he can cover all of the sites, one for each of three clients, on one trip. How should he arrange the billing? There are a number of possible methods. One is to share the transportation and travel time among the three clients; another is to bill the total travel time to each client and to bill each client for a full flight; and a third might be to share the travel time according to the value of the contract. Which ways are ethical and which are unethical and why?

The design engineer has another ethical dilemma under certain circumstances. She may be building the sample product for the customer or the actual product for the customer if she is designing and building a piece of process equipment. In the case of civil engineers, she may be hiring the construction crew, ordering material, and incurring expenses of a substantial amount if she is responsible for constructing the building or project. In such cases the original proposal will often have a rider stating that out-of-pocket expenses for materials, labor, and so on will be billed at cost plus 10 percent or more. Since this is part of the original proposal and not hidden from the customer there is nothing unethical about adding on a percentage of any value.

On the other hand, what if a design engineer is not buying large sums of material and not hiring large volumes of labor? Is it legitimate or ethical to add 10 percent or any other percentage to the cost of travel and other expenditures? Is such a charge justified? Some contracts for consulting work carry a rider that bills are rendered and due monthly, and that if bills not paid in 30 days, they are subject to an increase of 1-1/2 percent per month. Provided that this is agreed to by both parties, it is not unethical.

In one instance an associate of mine was called in for a deposition relating to an injury case. The notice to the law firm by the consulting organization stipulated that the consulting organization's minimum fee for each appearance by an individual was equivalent to the fee for four hours. This included travel time. The law firm called more than one person in for the same period. The second person arrived late. The consulting firm had advised the lawyers that this would occur. By the time the second person arrived it was evident that the deposition of the first witness would take at least the rest of the day. The lawyer therefore sent the second person away. The attorney later objected to being billed for the four hours. This was referred to a judge during a settlement and scheduling conference. The judge asked for the papers and was assured by the objecting lawyer that he had been advised of these charges before the event. The judge directed the lawyer to pay the bill. Do you have any comments about the question of the ethical behavior of either party in this case?

These questions of ethical behavior and billing relate to an incident that occurred about 15 years ago. The engineer was engaged in an investigation. One of the tasks involved the calculation of some 20 values which could be done using a table of logarithms or possibly with the use of a computer or a programmable calculator. The calculation using logarithms would certainly have taken several days, perhaps a week or more. It was suspected that after the calculations were completed the numbers in the table would change and the calculations would have to be repeated. This was all before computers. The engineer bought a programmable calculator, learned to use it, and charged the client for the calculator and the learning time. This billing was much less than had the calculations been done by the other available means, the log tables. There were no complaints since the client was billed for an equivalent time. The numbers were changed several times after the original calculation was made. The recalculation took minutes. The longest time was for writing down the answers. The calculator remained the engineer's property. Was this ethical? Subsequent clients were not billed for the calculator or the learning time. Was it proper that they get the advantage of the expenditure of someone else?

The questions of charges, use of design material on more than one job, and purchasing equipment that is used on the job or purchased for one job and then used on subsequent jobs have been discussed. There are other ethical problems relating to engineering, the design of equipment, and the supplying of engineering time.

If an engineer is traveling on an expense account for her company and she accepts a lunch and/or a dinner from the customer or vendor is it proper to list this lunch and dinner on her expense account? If you were in private practice and you were so treated would you list these expense items on your expense records that are used to justify the deductions you make on your income tax? What would you do if you were given a per diem? Suppose that the per diem was so low that it really never covered the cost of meals. The company would not reimburse you when you ran over and you were not expected to return money when you underspent. Is there an ethical way to report these expenditures on your tax return? In one company you were limited to $3.00 for breakfast, $5.00 for lunch and $12.50 for supper. The cost of meals regularly exceeded these figures. The answer of the corporation's treasurer was that you had to eat anyway, so that although this did not cover all your expenses you would have had to pay something when you ate, even if you ate at home. If you were on a trip and a meal cost less than the stipulated amount, would you report the lesser or the standard amount? Why would you take this action? Would you consider the company's action ethical?

12

Ethics Of Quality Control

Ethical behavior indicates that there are responsibilities which become necessary parts of the decision-making process of anyone who agrees to be governed by a code of ethics. It is possible that some people will not agree with specific aspects of some codes. It behooves such a person not to agree to the codes.

In today's industrial and government environment, the characteristics of parts, components, assemblies, and products are specified in a variety of ways, often extremely intricate. Some of the tolerances are specified with procedures or standard methods of measuring performance and conformance.

The task of quality control personnel will vary with their hierarchical position in the organization. The tasks may include:

- Examining, inspecting, or testing units for conformance to specification.
- Selecting systems, schemes, or plans by which a product or a sample of a product will be examined to determined whether a lot is in conformance with specifications, agreements, or standards.
- Selecting and designing systems for the statistical control of processes.
- Selecting and designing systems and plans for industrial experimentation to determine how to manufacture and produce units or products which are superior, or more uniform.

- Determining whether the specifications, contracts, or agreements properly delineate what is wanted and needed.
- Determining whether the process is capable of producing products, components, or assemblies which will be capable of satisfying the specifications, contracts, or agreements without sorting or reworking.
- Designing the controls for such a system.
- Determining how to control a process which will yield satisfactory material.
- Designing or participating in the design of an economically advantageous system of manufacturing control and process and product acceptance.
- Devising specifications which will adequately define product and delineate limits so that conforming units will operate satisfactorily and safely.

There are more limitations on some quality control and assurance tasks than on others. There may be several layers of quality control supervision, each with a different responsibility. The top echelon person is the head of an organization who may be involved in setting down the wide parameters of quality. These various decisions may include: all product shall exceed what the industry has accepted as standard; the product shall be the least expensive possible and must meet the industry standard, because otherwise it is totally unsellable; or the cheapest product possible will be made with a descriptive label even though the product does not meet the standards set by industry, which are often voluntary rather than obligatory. Some products bear the label stating that this is the company's best quality and conforms to their standards. Some very large companies have built their business on just this philosophy. Is this an ethical practice? Why or why not?

There appears to be another philosophy that is espoused by some managements. If they can sell the product there will be so much business from the sale of replacement parts that they will have a continuing market for their products. Does this philosophy, of living off replacement parts, override the philosophy that the product ought to be reliable and long-lived? Is it ethical to sell a device like a safety razor where replacement sales over a period of years will exceed the initial sale? In this throwaway society, is it ethical to make the parts that are thrown away larger and more expensive?

Another level of quality responsibility is that of the senior quality manager in a large company. The individual in this position has a variety of responsibilities some of which compete with the others. He or she may be able to sell the CEO the quality message that bad quality is more costly than good quality and that neglecting improvement is the most expensive course. Although better quality can cost less there are methods of trying to achieve better quality that are costly and wasteful. Eventually someone finds a different way to achieve better quality at lower cost. Decreasing process variability is one way this is frequently achieved. In fact a properly centered process with little variation, when compared with the tolerance spread, is much more effective than 100 percent inspection or any adjusting procedure.

Take each of the tasks listed earlier and ask yourself:

- Is this task being honestly and ethically done in my organization by me or someone in my department?
- Is this a task that should be done in our quality department or does it belong elsewhere in the organization?
- Do I or the appropriate members of our quality department become involved in the product at the appropriate stage of the development, such as initial proposal, development, release to production, manufacturing, sales, shipping, and customer complaint? If not, why not? Should we be involved at other stages of the product life cycle?
- Are we equipped to handle the responsibility and to provide input for the product at a time when it will be beneficial to the product and the process?
- Is there evidence that the product can be made within tolerance and with close to 100 percent first-run yield? If not, are there changes that would improve the probability of these conditions being achieved?

One common problem in an organization is that both the engineering and manufacturing departments look at the process design as their responsibility. Therefore, they do not like to be joined, in the process design control, by a third organization, the quality department. This leaves the quality department, which might have the greatest statistical know-how, in the unenviable position of trying to tell others how they should manage what they feel is their business.

The Process Capability Index is a measure that indicates the fraction of nonconforming product likely to be manufactured. The quality group ought to exert every effort to improve this index to a point where the fraction of nonconforming product is insignificantly small. Are you equipped to assist in this effort? Is this an area where the quality department is able to lend valuable assistance?

In one instance a suggestion by a quality control supervisor was resented so deeply that the two other departments complained that the quality person was infringing on their turf. The vice president of the division turned aside these complaints, saying that their resentment was due to the fact that the suggestion was good, and they knew that they should have thought of the idea themselves at an earlier date.

All of these concepts represent items which you should question. Were they ethical or what would be the ethical procedure should a similar circumstance occur in your company?

At all levels quality assurance and quality control people are interested in data which indicates whether an item or material conforms. Does it conform to the needs of the customer? Will it meet the customer's expectations? Will it continue to provide safe and satisfactory service?

Pressure from supervisors can cause inspectors to flinch, either to call a marginal unit good or bad depending on previous reactions from the supervisors and the situation. Do you call a marginal unit good when the lot acceptance limit has been reached? Or do you call it unacceptable? Under what conditions would you or your inspector make the decision one way or the other, assuming that this marginal unit causes acceptance or nonacceptance of the lot? Under what conditions should the quality unit allow the failure of a lot to cause a plant shutdown? On the other hand, suppose the lot had been manufactured internally and the failure to accept would mean missing a delivery. Would there be pressures to accept and call the marginal unit satisfactory? If you decided to accept the lot, would you recommend that the marginal unit be included in the delivery? In some instances a nonconformity can represent a serious hazard. If this were the case would your decision be different? Think of the many ramifications of this situation!

The function of the quality control inspector is often restricted to identifying conforming and nonconforming material, conditions, or process procedures. The presence or the absence of some characteristics or the presence of some dimension that does not conform to the specification may seem easy to detect. Sometimes it is simple and sometimes difficult. When dimensions on an item are close to the specification limit, it is sometimes impossible to decide whether the dimension is on the limit, over it, or under the limit. When the limit and the dimension coincide, what decision should be made? Some production people will insist that the piece conforms and some inspection and procurement people will want to judge the piece as nonconforming. The decision will be reinforced in one direction or the other by other factors or pressures. When the distribution of a characteristic is not centered or is wider than the tolerance limits, or whenever there is a concentration of many units at or close to the acceptance limits, this becomes a very serious problem. Some have said that the inspector's judgment is seldom questioned when a unit is accepted. Is it ethical to question an acceptance or rejection?

On one production line there was a final test for sensitivity on a phonograph pickup cartridge. A histogram of the output of these cartridges was made from the data recorded by an inspector at the end of the production line.

The histogram shown in Figure 12.1 has a vacant cell just to the left of the minimum acceptable output level. The inspection operation was observed by a supervisor after the histogram was drawn. The inspector had learned that when a cartridge was tested it did not always give the same output. Therefore when a cartridge gave an output just below the acceptance limit, the inspector would make

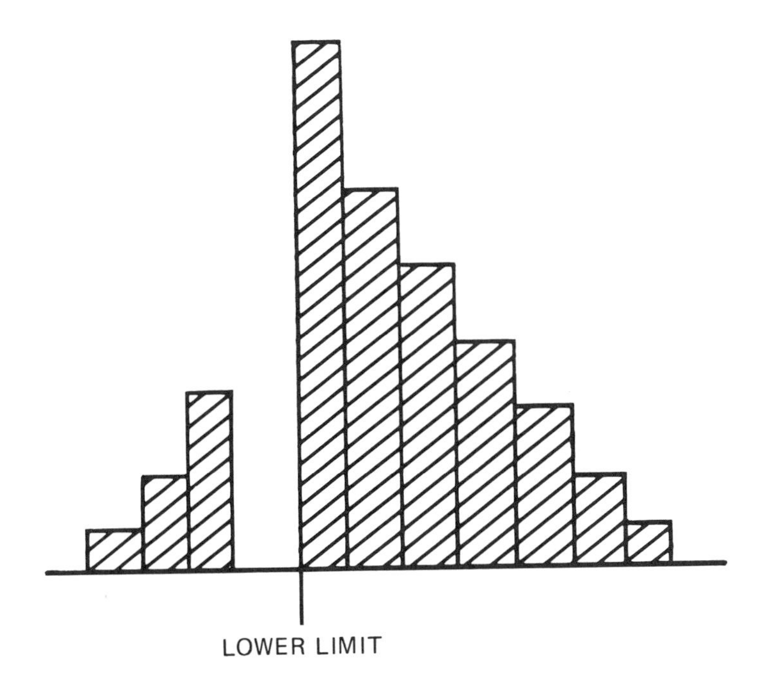

Figure 12.1. Histogram showing result of flinching.

a retest. The test was one that took only a few seconds. The retest might give a slightly lower or higher output. If the output was higher and the unit passed, the value was recorded and the unit accepted. If the output was lower on the repeat test, the lower value was recorded and the unit sorted into the unacceptable group. As a result of this procedure there was no record of any with output within the group just below the acceptable level, but there were many more than expected in the group just above the acceptable level. This type of flinching was acceptable to the inspector since it increased the yield and created less friction with the production manager. This was a strange situation in many ways. The customer had insisted on an output that was well above the minimum of the natural process and thus paid a premium for the units. The units with less output were more compliant and flexible. Therefore they had better tracking and high frequency performance and were readily disposed of to other customers who paid even more than the prime customer. The flinching satisfied the production manager but actually needed correction, because the low output units were salable at a better price. What is your opinion of the ethical behavior of the design, manufacturing, and inspection engineers?

The question of flinching is one akin to dishonesty. Inspection systems and inspectors sometimes look for ways to refuse acceptance of a unit or lot for reasons that are hard to ascertain. There can be pressures from some part of the management to accept or not accept shipments depending on the need for the product and whether there is some favored or preferred source. These pressures cause inspectors to reject what might be usable product. In this sense inspectors are looking for nonconformities. There are a wide variety of nonconformities. Any variation from the specifications is a nonconformity. The delivery of clear pine shelving when #2 grade shelving was ordered is a failure to conform to the specification. This deviation, the delivery of a better product than one ordered with no increase in price, represents the delivery of nonconforming product. However few will complain of this deviation. In general, inspectors do not have the authority to accept deviations from the specifications even when they are in favor of their employer. There are many nonconformities which are not of consequence. The delivery of an automobile with an upgrade in upholstery or the inclusion of some extras, provided the price does not increase, is another example. Sometimes even with a price increase the customer is willing to accept the inclusion of the extras, which in themselves represent nonconformance with the original agreement.

There is a group of nonconformities which represent serious threats to the welfare of users and bystanders. These nonconformities are called *defects*, and they not only can cause injury but may also result in the manufacturers, designers, or sellers being sued under the product liability laws. There are also a class of defects called *design defects* which can be responsible for customer dissatisfaction, loss, injury, or death. Despite the fact that all of the product conforms to the design, the product is faulty and not properly designed. Failure to observe safety and good design practices can result in each of the items in the entire production lot being defective in the sense that each item represents a serious hazard. The presence of manufacturing or design defects can also result in a recall, a very expensive process. Quality people may be and should be involved in the process of design review. This is more than the inspection of parts, assemblies, and systems. The inspection process is responsible for identifying nonconformities. The decision to classify a nonconformity as a manufacturing defect, or something that would be hazardous or unsafe, is a decision for engineers and supervisors to make in a material review after a nonconformity has been identified.

Among the most insidious nonconformities and defects are those that are the result of design deficiencies. The design review is a procedure with many objectives; one specific purpose is preventing or minimizing the likelihood of such an occurrence. The earliest review ought to be held before the product is designed. It is then that the designers are most flexible. It is also the time for others to provide input so that the design has the necessary features and proper form. This way the product can be safe and readily manufactured, and it will be free of hazardous and deleterious features. Quality people might be said to have an ethical responsibility to see that such design reviews are held not only prior to inception but also along the way. Does your activity include participation in such design reviews or does your first introduction to the product occur when it is introduced to production? What are the advantages of early familiarity with the planned product? Does such early familiarity lead to better customer service?

There is another form of inspection activity that is really dishonest. This is more likely to occur when material is passing from one entity to another upon the approval of resident inspectors. The dishonesty may be on the part of the inspectors, after all some people are corruptible; but it is more likely to be attempted by agents of the manufacturer or distributor, to hoodwink inspectors into believing that they are inspecting product when in reality they are not. It seems hardly proper to even discuss actions of this type under the heading of ethics, yet some may find some excuse or rationale to consider such action ethical under special conditions. Can you invent a set of circumstances in which you would consider taking such action and believe it to be ethical?

Some years ago a producer had a contract to make large overhead conductors for the Hoover Dam tie line that furnished power to the city of Los Angeles. The producer was but one of several holding similar contracts. One of the tests required by the specification was for corona, an electrical discharge from the conductor. All the other producers had trouble with this test. They all had electrical discharge currents which ionized the surrounding air and caused the cable to glow, of course, most visibly in the dark. Corona is an undesirable characteristic, but there seemed no way to prevent this occurrence under very high voltage test conditions. At lower voltages corona does not occur. One manufacturer's product consistently passed and showed no indication of the corona phenomenon. One day late in the fall, a testing laboratory's engineer who was acting as the customer's inspector worked later than usual. In addition daylight saving had just been discontinued. As a result the plant was dark when the high-voltage corona test was performed. The test bench indicated that everything was okay. However, overhead where the high voltage was fed to the test bench, there was a glow along the conductors up to a point where the test bench was spliced in. In the test area there was no sign of corona on the sample supposedly undergoing test. An investigation was made to determine the cause of corona up to one point and then a sudden cessation of the discharge glow. When the test setup was investigated, it was found that there was no high voltage applied to the specimen. What looked like an electrical connection to the high voltage source was an empty rubber tube rather than an insulated wire. In addition, the test bench had been rigged to provide false readings on the meters to make them look like they were actually reading the expected voltages. The inspector from the test laboratory had been faithful in reporting what he had observed, but the observations had been fraudulently produced. The test engineer immediately reported the findings to his supervisors. One must remember that, because of the lethal nature of the high voltage, no one touches or comes close to a sample undergoing a test of this nature while it is in process. Was the test engineer ethical in all of his actions?

In large cities, the practice is for the city government to provide building inspectors who pass on the adequacy of construction. There are two types of improper practice. One is to approve inadequate work or accept work that is not in accord with the building code, provided that a bribe is paid to the inspector. Another practice is to fail to accept construction work that is in accord with the code unless a fee is paid to the inspector. The payment is illegally made to expedite the work of inspection. If the bribe is not paid, the certificates of acceptance are not issued or may be delayed.

Inspectors of restaurants and food-handling facilities have recently been charged with similar practices in New York City. The practice was not confined to a few inspectors, but became so widespread that the city found it necessary to discontinue the practice of inspecting restaurants and food-dispensing sites [30]. The procedure here was to suggest that there were infractions of the law and that these would result in fines, which inspectors could take care of for owners. The cooperating establishments were not listed and the fine found its way into the inspectors' pockets. The *New York Times* of Saturday March 26, 1988 reported that the city of New York was introducing an undercover inspection team to identify and expose the malpractice of field inspectors [31].

The payment of gratuities and support for resident inspectors working in the plant are frowned upon. In most situations there is no minimum gratuity or gift the inspector can accept. In some instances such payments are offered and accepted. Are there gratuities that you deem acceptable? What are the limits of acceptable gratuities?

The Beech-Nut company was convicted of selling a colored sugar water with a little flavoring as a baby food labeled as apple juice. The mislabeling is a violation of the law. The cost of the product, the apple juice in a baby food jar, cannot be large, but when the quantity of product climbs into the millions the resulting profit can be significant. No one was poisoned and the product was not detrimental to children. Still it certainly is unethical. Suppose it could be proven that the product was in reality better for babies than the apple juice? Would this have been ethical? (Also see Chapters 4 and 6.)

When ethical behavior is discussed, so must the term *flinching*. Is flinching ethical even when it is known that the product just outside of the specification limit is acceptable insofar as performance is concerned? Is flinching acceptable if it is known that the product is superior to that within the specification?

Is it ethical to allow even a small percentage of the product to escape if that nonconforming product is likely to injure people or their property? Suppose that instead of the injury and loss being likely, there is a very small possibility that someone will be injured because of the performance of the product? Would you approve of such flinching? Where should the limit on flinching be set in this type of circumstance?

There are some products which can cause severe injury when used carelessly and improperly. These include knives, power saws, power equipment, and automobiles. Improper care while using matches can cause the destruction of buildings and death. The designers and manufacturers of these products must provide proper warnings and reasonable safeguards. Despite these practices there are many product liability suits for injuries resulting from accidents associated with some of these devices.

There has been established, through statistical analysis, a relationship between smoking and lung and heart diseases. There have been several suits, by heirs of those who have died from these diseases, to recover from the tobacco companies. To date the success rate for the heirs has not been good. The tobacco companies have sustained that they were not guilty of any wrongdoing or carelessness. One recent suit had an opposite outcome.

A similar attitude was originally taken by the manufacturers of asbestos products. The asbestos people were not as successful as the tobacco companies. The courts have upheld the claims of those suffering from various asbestos-caused diseases and the claims of those who are their heirs.

The manufacturers of the Dalkon Shield, an intra-uterine device intended to prevent conception, have also been unable to sustain the claim that they did not have knowledge of adverse effects and that they had properly warned the users of the dangers inherent in the device. These lawsuits run into astronomical dollar amounts. One maker of drugs has estimated that if it were to lose the product liability suits that have been instituted against it, and the courts were to award the amounts claimed by the plaintiffs, it would use up 400 years of profit just to pay for the suits.

Paying for the losses due to the adverse effects of drugs makes it too expensive to produce and introduce some drugs; thus many drugs are never introduced, and drug companies will not work on medications for diseases that are not common to large population groups. The cost of qualifying drugs so that they can be sold is not worthwhile as far as manufacturers are concerned. In addition the risks of the losses from those that were injured by the drugs are greater than the profits that might be made. Ethically should there be a way for the government or a consortium to underwrite these losses or provide a method of underwriting them?

Should there be a limit on claims for injury from the side effects of drugs? If the manufacturer is aware of the specific adverse effect and does not warn the physician and the user, would your opinion be the same? In answering this question you must remember that there are a few individuals who are going to be injured by the drug even though it is strictly in compliance with the specification. It will also be extremely useful to others. You must also remember that only a short time back all children were vaccinated against smallpox. Smallpox is for all intents extinct. Vaccination would severely injure or kill more children than would be saved by the procedure as of this time. Years back the reverse was true. When does it become ethically proper to discontinue the use of a preventive procedure?

There may be times when individuals, be they engineers, executives, or inspectors, do not believe that their company is operating in a manner that serves the public safely and properly. These workers must question what they are doing to maintain their ethical position. There may be several ways. One is to talk to superiors. Another is to discuss with fellow employees, and still another is to go to the head of the company. Perhaps none of these proves satisfactory, and the individuals must decide whether to separate themselves from the operation, resign, or go public with the information and blow the whistle on the unethical or dishonest operation. They must recognize that whistleblowing is hazardous; perhaps even more hazardous than resigning, as can be seen from the discussion in Chapter 9.

In an effort to obtain factual information on the ethical behavior of engineers, quality control personnel, and others, letters were sent to the ethics committee of the Institute of Electrical and Electronic Engineers, The American Society for Quality Control, the American Society of Civil Engineers, and the American Society of Mechanical Engineers. In the following discussion these societies will be referred to by their initials, IEEE, ASQC, ASCE, and ASME, respectively.

Information as of this writing was obtained from several of these societies. Naturally all of the societies took the position that the information relating to any one case could only be discussed with the name of the individual purged from the discussion.

The largest of these societies, the IEEE, stated that there was no record of any action by the ethics committee that any member had been censured, suspended, or expelled.

The ASME reported that my letter was referred to the ethics committee for a reply. Perhaps this society is more sensitive to the question of its actions because of its bad experience in the Hydrolevel case discussed in Chapter 3.

The ASQC could not find or recall the censure, suspension, or expulsion of any member. The ethics committee members recalled several cases which had arisen, but there had been no action taken to date to censure, suspend, or expel a member.

The problems that had been called to the attention of the ASQC ethics committee were interesting and informative. They are indicative of the questions that come up and you should review what you think the proper action would be in light of the ASQC Code of Ethics.

There have been cases where individuals in the society have passed on to members and individuals in foreign countries material that the society felt was confidential. The society thought that the material should not have been given to the foreign individuals without consultation. These cases were settled with agreement that the practice would be discontinued. There was no mention of exactly what material was released. If it had been material that was covered by a military classification, this would obviously have been a different matter and would have been reviewed by the military and the courts.

There are situations where a member of a section or a conference committee used large sums of money for his or her own purpose and at least one instance of defalcation. Once the question of ethics was raised when a conference program chairman promoted the event by using the names of several important public figures and indicated that they would address the conference. Upon opening the meeting an announcement was made that these several people were unable to appear. Actually none of the individuals had ever been contacted for this program.

The ASQC ethics committee considered whether it would be proper to allow inmates of a prison to become members of the society. There is interest in quality control in some of the nation's prisons where there are operations such as the manufacture of automobile license plates. Would you consider it ethical to permit prison inmates to become members of the society?

There were other items that were referred to the committee. Is it proper to indicate in an advertisement that the consultant received a medal or award? Is it ethical to indicate that the award had more prestige than it does in the view of the society management? Is it proper to sign one's name with a title? Is it proper to use the term M.D., P.E., or Esq., indicating the fact that the individual is a physician, professional engineer, or an attorney? Would it be proper to follow your name with FASQC; Fel, ASQC; or Fellow ASQC? If it is proper, are there any limiting circumstances?

Earlier in this book the case of an ASQC member using the term *quality control* as part of his company name was discussed. This issue never came to a vote as there were others who believed that the term is generic and no more a property of the society than the term *electrical* or *mechanical* are the properties of the IEEE and the ASME.

A more serious problem that may be facing the ethics committee is whether it is proper to use the society logo when making bids on projects. Do you believe that this would be an ethical practice?

There is also the question of whether nonmembers who become certified quality engineers are bound by the ASQC code of ethics. When should a member be given a copy of the ASQC Code of Ethics and be bound by it? In some instances the Code of Ethics was printed on the application for membership. There are some membership applications which do not have the code of ethics on them. In the latter case is the new member bound by the code?

Of all the replies that were received, the most voluminous was that from the American Society of Civil Engineers. As noted the IEEE had no record of any activity in the ethics committee, but in contrast the ASCE had more than 500 incidents in the 15 years of record furnished. Approximately half of these cases have been continued. More than 20 members were expelled or allowed to resign with prejudice. Approximately 40 were admonished or suspended for periods of one to five years.

It is obvious from a review of the cases that civil engineers operate in a different sphere than most other engineers. They are engaged in public works and in some cases are state, town, or county engineers and represent the interest of a government entity. In some instances there is a conflict of interests. Individuals had been or still are members of firms interested in obtaining contracts for jobs. Civil engineers in many instances are in competition with others in a much more direct manner than other engineers. The civil engineer is the principal in many civil engineering firms.

The ASCE expulsion, suspension, and censure actions have been the result of members being improperly involved in kickbacks, political contributions, bribery, conspiracy to defraud, effort to supplant, extortion, plagiarism, lying, falsification of records, injuring the professional reputation of another engineer, use of confidential information for professional gain, bid rigging, incompetent preparation as an expert witness, and failure to hold paramount public safety, health, and welfare.

The list of misdeeds is impressive, and it is easy to see that other professionals can become involved in many activities that do not differ from those listed. An individual nonmember might be involved in activities that are, on the surface, unethical.

It certainly behooves all of us to look over this list of activities and ask whether some of the activities of members of our organizations are not similar to those listed.

The ASCE has also admonished members for a variety of other activities. These include filing false federal corporate income tax returns, soliciting campaign funds or political contributions from employees while a county engineer, making interest-free loans to an administrator of a government in a conspiracy to falsify time sheets and other records while employed as a chief structural engineer. This resulted in excessive charges to the county.

The ASCE has also made available a number of detailed Professional Conduct Case Studies. There is one in which a professor is accused of plagiarism. It is charged that the professor published a paper under his name without any credit being given to the author, one of his students. There are arguments on both sides of the case and a statement of what the known facts were and how the case was resolved. The other cases covered are treated in a similar manner. There are 14 such case studies.

There are many situations in which quality control personnel could become involved in ethical misbehavior. There are undoubtedly some which have occurred to each and every one of us. Perhaps the situation was not very serious. Perhaps it was like little white lies.

As examples of the questionable acts that quality control professionals may perform, consider a lot of material that is undergoing attribute sampling. The sample is randomly chosen and the lot meets the acceptance criteria. While placing the conforming sample items back into the lot the inspector notices several obviously nonconforming items in the remainder of the lot. Should these items be removed, discarded, or replaced? It should be noted that MIL-STD-105D stipulates that the supplier does not have the right to supply knowingly any defective (In MIL-STD-105E the word "knowingly" is omitted.) unit of product. In ANSI/ASQC Z1.4 the reference is to a nonconforming unit.

Suppose that the lot contained 3500 pieces and was being inspected to an AQL of 0.04 percent (400ppm) using general inspection level II and normal inspection. Further assume that there has been a series of lots from the same source, and they have been inspected in accordance with the proper procedure, and that the switching rules have been properly applied. This lot requires a single sample of 315 with an acceptance number of 0. While replacing the sample into the lot the inspector notes two units which do not conform. What is the ethical action? Should he or she remove the two nonconforming units and ship the lot? Should a supervisor be contacted? Note that two nonconforming units in 3,500 exceeds an 0.04 percent quality level.

The MIL-STD-105D sampling standard and its industrial and international cousins are among the most misused specifications of any I know. Not only do many corporations and their employees misuse them, but government agencies do also. Many of the organizations fail to use the switching rules and many use improper sample sizes and acceptance numbers. There is a firm belief that if a lot fails sampling, 100 percent inspection will remove all nonconforming product, and few of these organizations bother to resample the lot after sorting out nonconformities. Is it ethical to use only that portion of a standard that one feels like using and ignore the balance?

There have been reports of government contracts administered under conditions where sampling was used, and if the number of nonconforming units in the sample exceeded the allowable number, these were removed and the lot shipped. The organization said it had performed the sampling and removed the nonconforming units. Is this ethical?

It was reported by a quality engineer that the following conditions prevailed in the plant in which he worked. He requested guidance. At the time this was written there was no way to verify the accuracy of the statements, but they are nonetheless interesting because they indicate how much confusion can occur in attempting to apply MIL-STD-105D.

The engineer's (edited) comments concerning clauses 1.4 and 2.2 are included. He quoted from these clauses and then stated what he believed the practices were in the operation with which he was associated. I am not familiar with the operation and cannot comment on the correctness of the observations. You might ask whether the described organization is performing the acceptance procedure properly or whether there is some interpretation of the specification not readily available to the engineer.

> *1.4 INSPECTION BY ATTRIBUTES* Inspection by attributes is inspection whereby either the unit of product is classified simply as defective or nondefective, or the number of defects in the unit of product is counted, with respect to a given requirement or set of requirements.
>
> COMMENTS *In our operation, under percent defective inspection, the unit must be classified as defective if it contains one or more defects. This is not done. Furthermore, each defect on each unit of product is listed and counted, but no lot rejection will take place unless four or more defects are found having the same defect characteristic. This does not seem to agree with the requirements of clause 1.4.*
>
> 2.2 *MAJOR DEFECTIVE* A major defective contains one or more major defects, and may also contain minor defects, but contains no critical defect.
>
> COMMENTS *Any unit containing one or more major defect—but containing no critical defect—must be counted as a major defective for acceptance and rejection purposes. This is not done in our operation. As many major defectives may be found as there are characteristics in the unit of product that are classified as major, and the lot will be accepted unless four major defects of the same characteristic are found. There are 29 major defect characteristics* [32].

Other areas where unethical behavior can occur is in the classification of an article as conforming or nonconforming when it is borderline. There is never much difficulty in deciding that an article which is midway between the specification limits is conforming or one that is well outside one of the specification limits is nonconforming. When the distribution overlaps one or more limits, the action often taken is to get more expensive instrumentation to assure that the limit is more accurately defined. This makes for fewer marginal decisions. Another way is to retest all marginal failures. Is this ethical? A way this can be avoided is to improve the process so that the process variability is markedly less than the distance between the tolerances. Is one action more ethical than the other?

It is well known that when there is a large density of units at a specification limit, some nonconforming units will be classed as conforming. The reverse is also true, some conforming units will be classed as nonconforming. If, as is commonly the case, the average of the distribution is within the tolerance values there will be a slightly higher concentration of units inside the tolerance limit than just outside. As a result there will, on the average, be more good units classed as nonconforming than nonconforming classed as conforming. Manufacturing supervisors have been known to review the units classed as nonconforming and reclassified some. Is this ethical? Does an AQL widen the tolerance limits by permitting the acceptance of lots containing a small fraction outside of the tolerance? What do you believe is the ethical approach for a quality supervisor?

In some organizations the factory limits are purposely placed closer to the average than the true tolerance. This effort would eliminate marginal units. If the manufacturing capability is to produce a distribution that is narrower than the tolerance limits, then this is usually very successful. If the distribution overlaps the tolerances there are organizations that place the factory limits outside of the tolerance limits. Would you consider this ethical?

Grubbs and Coon discuss this issue in detail from a statistical point of view. They claim that the most advantageous procedure is to set the test limits outside of the specification limits when there is measurement error.

Many variations to these situations can exist: the distribution can be normal and well centered; it can be a multiple distribution centered on a tolerance; and it can be a rectangular distribution. These factors can alter the situation so that the best decision in one case is not the best in the other. Søren Bisgard of the University of Wisconsin Center for Quality and Productivity Improvement at Madison discussed some of these variations in "Design of Standards" delivered at the August 1987 ASA Annual Convention in San Francisco. He too found that, with the usual distributions and the median well within the specifications, it is advantageous and most satisfactory from the point of view of the supplier and the customer to test to limits that are outside rather than inside the specifications.

When discussing the ethics of quality control, you may be hard pressed to define the ethical procedures that you ought to adopt in each and every instance. What would you list as the prime objectives of quality control, quality assurance, and quality systems?

Some might have a different point of view, but the concept of most major systems users is that the purpose of quality systems is to economically reduce the cost of producing conforming product and to assure that conforming product is shipped on a timely basis.

In many instances there is a major interest in not letting a nonconforming unit get out the door if the nonconformity is of a type that could adversely affect use or the safety of users or bystanders.

The simplest form of quality control operation has an ethical duty to reject nonconforming material and to accept conforming material. This is an easy assignment when there is nothing at the borderline. When there is a lot of material at the borderline this is more difficult, and for practical purposes, it is impossible. Take the simple case of a military shell that must fit into a rifle chamber. If it is too large it will not fit. There must be a maximum size that is allowed and nothing larger. There are many procedures that can be and are adopted to assure that nothing oversize enters the final product that is shipped.

As was previously noted, there are many situations where there is a difference between a nonconformity and a defect. A nonconformity is a failure to conform to specifications. The inspector is charged with identifying whether or not the product conforms to the specification. Sometimes there is no problem associated with a failure to conform to specification. In some products there are internal plates or chassis on which parts are mounted. Some of the holes punched in these plates are ventilation or clearance holes. If they are oversize by a small amount there is no real problem. The plate is not weakened sufficiently to affect its usefulness and the extra clearance has no ill effect. What is the most effective way to handle this problem? Is this the most ethical way?

There are other instances where failure to adhere to specifications is of no real consequence, as in the instance previously cited where a lumber yard delivered, at no extra charge, a quantity of clear pine shelving rather than the ordered #2 grade which allows for the presence of some knots. In another case a paint was delivered that did not match the ordered sample and was not within the specified color limits. Since this paint was used for an interior portion of the product, it was not considered unsatisfactory even though it did not conform. In another situation the paint was to be used for panels that were to be assembled with previously painted panels. Since the paints did not match and the differences were such that the customer would complain, the shipment was defective, even though the paint was perfectly good. Appearance standards might even suppose that the paint was suddenly found to have a poor match under a certain kind of light. This was not in the specification. Is it proper to class it as nonconforming or defective? It may agree with the specification but be unusable. Ethically, can it be rejected? Here there may be a legal obligation to accept it. Even under these conditions it is possible that there might be a beneficial solution that can be worked out.

A product that is nonconforming is one which does not meet the specification. If this failure to meet the specification adversely affects the use, may cause the product to fail, have a short life, or cause danger to a user, a bystander, or property, then the product is defective as well as nonconforming.

Actually the short life or the potential danger may be the result of poor manufacturing, the purchase of poor material, or poor design. The fact that a product injures someone or destroys or damages property, assuming that the incident was not caused by user misuse and carelessness is almost always traceable to an aspect of quality control. This includes the failure of a quality control procedure, or a totally inadequate use of the quality function, or the use of neither inspection nor quality control.

To discuss this further, one other form of incident which results from the interface of the product and the user must be described. A match can cause a fire and burn down a building or ignite the clothing of a person and do serious injury. This is not necessarily the fault of the match. A circular saw can badly cut a user or remove a finger. This can occur despite many guards. If the purpose of a saw is to cut wood, the saw can cut other materials as well. Guards, good design, and safe handling practice can reduce the frequency and severity of accidents, but it is unlikely that there will ever be a complete elimination of saw-related injuries.

Lead-acid storage batteries used in automobiles are useful devices that make it possible to start the engines by turning a key. Ages ago when engines were smaller, and even today in the case of small engines such as used on lawn mowers, it is possible to start the engine with hand power. The old system used a crank, the newer one uses a lanyard that is pulled and spins the engine. Storage batteries are useful devices, but under some conditions they have exploded and done serious injury to owners or bystanders who often suffer eye injuries. Improved batteries have lessened the frequency of explosions. Based on the definitions, batteries are always dangerous products and some have been defective in design and some defective in manufacture. Some batteries, the size of a pill, have been ingested. All of these factors contribute to numerous injuries and claims. Are the warnings on these batteries adequate?

By definition a good quality system would eliminate or at least reduce the frequency of improperly manufactured product, and the product design would have been reviewed to assure that it was not defective. The quality system would also need to have a system of preventing the escape of defective product, particularly product that would cause injury and loss. Good operating procedure dictates that there be few losses since these can be very costly to the company. Defending a company in court is an expensive procedure, so is paying for the loss or injury. Arriving at a settlement may be less costly since a jury may award the full amount that is requested by the injured party. Quality managers and others must therefore try their best to see that the product, as manufactured, performs its intended function. Despite these efforts the case may wind up in court and among those called to testify may be quality managers. They may be asked to explain how each component, assembly, and final unit is tested. They may be asked if each unit is tested or if the lot is just sampled and why. They may also be asked to explain why the company is willing to test a sample of the units it makes, when any one of them, particularly if they are like the one the injured party had, could seriously injure a person? Do the quality managers have data to substantiate their statements? Some believe it is wiser to go into court and report that a company does not keep data. If the data shows that you have been careless and allowed the shipment of product when you found serious defects, and made no effort to search for more defects, if you continued to accept lots when some of them had defects, you may be better off without data. If you have been diligent and have records that show that, do you think that you ought to destroy them and have the jurors doubt your diligence and your ethical behavior? What is your opinion? Does your company have a policy? What does your record retention department recommend? What would your legal counsel recommend? Is there a difference? Why?

What has been said about quality managers might also apply to others in the company. If you were called into court would you all depend on the same data or would each have a different set? If there were inconsistencies, what would or might the jurors believe? Would it benefit your case?

In one case a major loss was sustained by a computer company. This case related to the use of the attribute sampling plans of MIL-STD-105D. Youcorp manufactured computers which sold for several hundred thousand dollars. Some capacitors which were used in the assembly of this computer were not of the correct value. This was first discovered months after many units had been shipped, when a user tried to exercise a seldom-used function. It cost Youcorp many thousands of dollars to identify the location of each of these off-value capacitors. It requested Capcorp, the capacitor manufacturer, to pay for this retrofit procedure.

Capcorp answered that it had shipped these capacitors at a 1 percent AQL as was common in the industry and that Youcorp had reported that the quality was better than this. Youcorp admitted that it had accepted all shipments on a sampling basis and that all had shown fewer nonconformities than were allowed by the 1 percent used as an acceptance criteria. Capcorp said that it saw no reason to offer Youcorp any payment.

Youcorp said that unless it received payment promptly Capcorp would receive no more orders. Capcorp countered that the threat was of no merit and that it had now made its last shipment to Youcorp. It must now consider all its orders canceled.

The case was then referred to attorneys for legal action. What are your opinions of the actions reported? Was it proper for Capcorp to ship these capacitors after the lots passed sampling to the 1 percent AQL? Was Youcorp ethical in requesting payment for the retrofit procedure? Was the contract or the shipment-and-acceptance procedure conducted on an ethical basis? If you were to now negotiate a contract of this type would you, as either a supplier or purchaser, recommend there be any changes from the acceptance procedure described? Are there ethical considerations either side should have considered in this case? If you were a quality control manager for either corporation would you consider that your organization was performing in an ethical manner?

The Society of Automotive Engineer's, (SAE) standard for automobile turn signal flashers, circa 1972, contained a life test that determined the suitability of flashers for service in the vehicles licensed for use on U.S. highways. The life test could be performed in approximately 200 hours and required 17 of the 20 units placed on test to survive.

The National Highway Traffic Safety Administration, NHTSA, in its effort to promulgate national rules for equipment that impact safety in motor vehicles rewrote the specification and published it in the Federal Register [33].

Like many other U.S. and foreign government ordinances NHTSA contends that sampling plans permitting any fraction of product that is nonconforming or any failure during a durability test would establish, "permissible failure rates and would therefore raise difficult problems of interpretation and enforcement, and in any event is not in accordance with the need for motor vehicle safety."

Within the agency, the engineering staff found that the rewriting of rules to permit no failures was impractical, given the states of the art of flasher development. They therefore revised the specification to broaden the performance deemed acceptable at the end of the durability test, materially shortened the durability test, and left out any reference to the number of units that were to be placed on the durability test. According to the new specification the life test could now be performed by testing a single unit.

Because of the aforementioned interpretation of standards and safety requirements, government agencies resort to testing individual items or so few samples that there is a high probability of obtaining 100 percent compliance with all characteristics including durability. The effect of this type of a procedure is to allow the durability tests to be very short and the samples tested very few, thereby opening the way for the introduction of devices, equipment, or flashers, with shorter lives. This can result in there being more failures on the highway and the need for more frequent replacement. NHTSA is charged with improving safety on the highway. The imposition of zero failure durability tests has the effect of allowing product with shorter service life to pass acceptability tests. Thus the zero-acceptance philosophy, rather than assuring better product, allows the development and acceptance of poorer product. Nonetheless the zero-acceptance philosophy has worked its way into the laws of many nations. Do you find this an acceptable situation? Do you have suggestions for corrective action to improve the rules?

These conditions make it necessary for the continuance of zero acceptance number plans in sampling tables. Therefore these plans are necessary in sampling tables such as typified by MIL-STD-105, ANSI/ASQC Z1.4, and ISO 2859.

These are legal and ethical questions and the problems which occur when law and science are not properly coordinated before laws are enacted. Man-made laws cannot be effectively enforced when the law is written in a way which contradicts the laws of nature. In one instance a state legislature passed a bill stating that the value of pi would be three.

Ethical quality control must therefore be involved: with the design review system to prevent the adoption of defective designs; in the manufacturing process to assure that there is an adequate control of the manufacturing processes; in purchasing to assure that the proper product is purchased and that the supplier is capable of supplying the material, component, or assembly in a manner that is consistent with an acceptable final product; with the sales, labeling, and packaging; and finally, in writing of legislation. Users must be apprised of all cautions and dangers, assuming that they cannot be eliminated, and there must be proper instructions, labeling, and packaging so that the product arrives at the user's site without damage. In some instance the product must be installed and the instructions must be adequate and the pitfalls eliminated.

Quality managers must be hardworking individuals to pay attention to all of the areas of care and see that the procedures that are used are correct and appropriate to the circumstances. Ethical service requires no less. It is therefore also the ethical duty of the quality practitioner to become educated in the intricacies of the quality system techniques and their limitations and constraints. To be conversant with the appropriate techniques for the operation is an ethical responsibility. Each of you should list the techniques appropriate for your operation and for other operations with which you are familiar. Are the personnel involved being ethical in maintaining their technical competence?

In most cases it is necessary to thoroughly understand how the product is supposed to work, how it really works, and why it fails, if and when it does fail. Would you consider engineers or quality control professionals ethical if they did not know how the product was made, why it functioned, and the necessary factors for it to function safely and to fail safely?

The concept of a fail-safe design is best illustrated by the old railroad air brakes. The railroads in their early days experienced frequent accidents. George Westinghouse invented the air brake. The brakes were applied by heavy springs. When a train was made up the cars were coupled together, but an air hose was also coupled that ran from the locomotive through each car. When air pressure was applied to the hose, the air cylinders in each brake would act to compress the spring and release the brake. If one or more cars separated from the train, the air hose would open and the air pressure would be lost. The result was that the springs applied the brake to the front end of the train and to the lost cars. Thus the train's brakes would automatically be applied in the event one or more cars became separated. Applying the brake by applying positive pressure would not accomplish this and the system would not be fail-safe. It would be ideal if a fail-safe solution for every condition could be devised.

In one plant an electronic assembly was manufactured. The integrity of the soldered connections was extremely important. The parts were small, and the joints were not easily made unless all of the parts that were to be soldered readily accepted solder. Properly cleaned and pretinned parts accepted solder, and the operators were able to see that the solder flowed properly and that a good joint was made. Due to the fact that it had been proven that it was important to see that the joints were properly made, the manufacturing staff had been instructed to halt operations if the solder did not flow well at the joint. The operators were, in fact, doing the inspection and were controlling their work. When a specific part did not solder properly the line foreman was notified. He would stop the work and see that the parts coming in were cleaned or pretinned ahead of the operation or recleaned and retinned as might be appropriate. This aspect of the quality control function was exercised by the manufacturing department without any friction. The concept was that the manufacturing department was to control the integrity of its own operation. With this type of system in operation it was found that defective solder joints could be reduced to approximately one in 50,000, or fewer. Since there were almost 100 joints per unit this meant that fewer than one in 500 assemblies were defective, or likely to give trouble for solder joint problems. This represented a vast improvement over the previous situation where the rate of defective joints had been found to be between one in a 100 and one in 1,000. Does the control of the process to the manufacturing department represent an abandonment of the duty of the quality control and inspection group?

There are often questions in the plant and in the laboratory concerning the effect of some changes. The simplest example is the question that faces everyone when there are two ways to do some operation or assembly. The experience in the plant is that some of the assemblies are not working properly. The percentage is small. Nonetheless the fact that there are some assemblies that are not working correctly gives rise to an opinion that a different way of making the assembly may result in a much larger percentage of the product working.

A suggestion is made of a way to change the procedure. Someone assembles 10 units and they all work splendidly. Has it now been proven that the change is effective? Not necessarily. Suppose that the percentage defective had been as high as 10 percent. The probability of getting zero or one nonconforming unit in 10 is essentially the same. Therefore it has not been proven that the procedure was better, and the chance of getting more than one in 10 is not much less than getting zero or one.

Suppose that the fraction percentage of nonconforming items had been smaller than 3 percent. Then the fact that 10 had been made with no nonconforming units proves even less. According to the binomial distribution, the chance of getting zero nonconforming items in a sample of ten from a process that is yielding 3 percent nonconforming is approximately 74 percent.

Should the changed conditions be given credit for resulting in the 0 percent nonconforming in the sample of 10? Ethical considerations indicate that the quality control statistician should give an opinion as to whether the data suggests that the manufacturing process has been improved. Should the probabilities that the process rather than being improved had worsened be explored? There are three possible outcomes when such an experiment is performed; the nonconforming fraction may increase, decrease, or remain unchanged. Is it not an ethical responsibility of the quality engineer to prevent people from jumping to an unreasonable conclusion?

Similar situations occur with great frequency. A lot fails acceptance and rather than being resorted is given some form of treatment. This does not change the appearance but a physical property is reported to be changed. If the tests are made by an attribute inspection, how different ought the number of nonconforming units be to provide proof that the lot ought to be accepted? Suppose the lot consisted of N pieces and n had been in the original sample. When this sample was inspected the acceptance number had been exceeded by one. After treatment the number of nonconforming units in a sample of the same size, n units, was equal to the acceptance number. Is there evidence that the treatment has improved the conformance of this characteristic of the product? Should the lot now be classed as acceptable?

This question of interpretation is one which requires a little statistical know-how. Ethically, how much statistical knowledge should quality control departments have in order to prevent gross errors from being made when changes in construction, design, or procedures are to be accepted? It would be catastrophic to make a large expenditure to correct and improve a situation and then discover that the change has resulted in a worsening of the process and/or the product. Is there need for a more sophisticated type of statistical experimentation or for more knowledgeable individuals to analyze the results and indicate what should be done?

There are other situations which arise in industry. Most of the investments in equipment are made to improve product and reduce costs. The following examples are indicative of things that can go wrong.

EXAMPLE 1. A production line had a stage in which a liquid was inserted into a can. The liquid was corrosive and when the liquid spilled over the edge of the can, the product had to be washed and cleaned following closure. This was no small operation. There was a continual flow of cans and the washing and cleaning took four operators on each of two shifts. A change in the filling technique which used a long needle inserted down near the bottom of the can resulted in no liquid flowing to the outside. The washing and cleaning operation was unneeded as was the machine, which had cost more than $10,000. The machine was declared unnecessary and offered for sale. The machine was less than six weeks old. Couldn't some use be found for it? The plant superintendent who had the machine designed and built to cut the washing and cleaning labor in half was no longer on the job. Why did the new management of the plant insist on scrapping $10,000? Remember the labor of eight employees was no longer needed and the cans were in better shape with no washing and cleaning. Was it ethically correct to criticize the new management for scrapping the machine?

EXAMPLE 2. A manufacturer of a product which had been approved for a military application noted that the product's performance was marginal on one test. They decided that rather than run the risk of items failing on test or in service, to make a change in the product. The proposed change had other ramifications. The product performed better under most of the other tests and service conditions and under no conditions did it perform less well than the original design. The difference between the changed and the original product could not be detected by appearance. The new design weighed a trifle more than the original but still was well within specifications. What was most favorable, as far as the manufacturer was concerned, was that the product in its changed version cost considerably less to produce. The manufacturer stood to make a substantial profit on the contract, whereas before the change the probability was high that the profit would be very small.

The product was changed. Eventually an investigating office discovered the differences between the two versions. It spent nine months examining costs and attempting to prove that the company had shortchanged the government, with the acquiescence of the purchasing agency and therefore was required to compensate the government,. The investigation failed to prove that the revised version was inferior to the approved version in any way. The fact that the product weighed more was taken to indicate that the government was getting its full share of product even if the material within the package was slightly different. After much negotiation the company offered to reimburse the government $1,000. This was accepted and the case was closed. One of the most pleased was the purchasing agency because it was now cleared.

This example has many facets and you might examine the ethical practice of the company, the procuring agency, and the investigating agency and the ethical practice of settling for $1,000. The cost to the company was much greater, because there had been four investigators on the premises for nine months each.

EXAMPLE 3. The manager of a production plant hired a firm to design and build a machine to perform an assembly task. The machine would do the job of 18 people and would require six to operate it. The savings would be significant because the operation was on two and a half shifts. It was calculated that the machine would pay for itself in 10 months. The delivery actually took about 18 months. In that period the operator efficiency had been improved by a minor change in the task. The operators were doing more than three times their former rate. The machine, if operated at capacity, would require more labor than the simple jigs that were currently in use. Again the financial management was aghast at the thought of scrapping a machine that had cost so much and had never been put to use. The new manager of the plant was charged with the scrap and roundly criticized for not making the machine work. The ethics and economics of the situation are worth discussing. Should the new manager have tried to salvage the machine and introduce improvements so that the machine would pay its way? Was the top management ethical in placing blame on the new management?

The legislatures of various states and other government agencies have been cited for hiring employees who do not do any work, who do not show up for work, who are used by incumbents to assist them in their political or private efforts. There are many situations where individuals who hold office, elective or appointive, are able to do their own work from the office supplied by the agency. In some instances the incumbents are more interested in representing and pursuing the interests of clients than the interests of the people. Legislators have been known to be in the employ of corporations and private parties and act in every way they can to do their best for these interests rather than the public good. There is no need to discuss the ethics of these situations, but it should be noted that in many instances the acts are not defined as illegal according to governmental codes. Nonetheless those who act in this manner do not readily disclose the arrangements.

In contrast, most industries frown on such activity. There are, however, many instances where there are people with multiple interests who steer a company's business in the direction where it will bring them the most profit.

In one instance a pharmacist ordered some products from a supplier who was partly owned by the pharmacist. The manager of the organization employing the pharmacist discovered the connection and fired the pharmacist. It was pointed out that the same supplier had been delivering these same items to the organization for several years prior to this pharmacist being hired. It was also highly probable that the organization was getting the best price it could. Was the action of this pharmacist ethical? Was it ethical for the manager to terminate the pharmacist? Was there some other solution or action that might have been taken before or after the discovery that would have been more ethical?

13

Ethics in Business

While it is the purpose of this chapter to talk about the ethics of business, it should be understood that this is in the interest of bringing the subjects of quality, business and ethics together. It may seem that some of the examples are more quality ethics and less business and some are just the opposite, more business and less quality, and some are rather general questions that relate to subjects that have been in the news recently.

The question of properly completing a task is a quality judgment. If the task is done to everyone's satisfaction, primarily that of the customer and the supplier, then both parties are satisfied. If the task or agreement is completed and both the supplier and the customer are satisfied, but there is a point in law or in public policy that has been circumvented, then the transaction may be illegal or unethical.

The simplest example of such an arrangement is the selling of crack cocaine. The dealer delivers and the customer gets good crack. The law prohibits the use, possession, and sale of crack. The same is true of marihuana. A patient in a hospital is suffering from cancer and has just had a treatment which results in her being in pain and upset. It is said that a short smoke of pot will give her major relief. Some kindly souls chip in, and the marihuana cigarette appears. It gives the woman much relief. Was this unethical?

A painter quotes a fixed price for decorating the interior of several rooms. After the job is completed he returns at the request of the owner to retouch several places where the work seems unsatisfactory. He does this with the exception of one room. In this room he states that it will be impossible to retouch because the paint did not seem to go on quite right. He will return with a helper in a week and redo the room, with no additional charge. Is it ethical to accept such an offer?

Most professionals quote their prices on the time used. Some few will quote on a job, and sometimes it is necessary to bid on a complete job. A surgeon will often advise a patient that an operation will cost a specific fee. There may be complications and more visits than anticipated, but just like the DRG's of Medicare and Medicaid, the average is well known to those in the field.

I find that I am sometimes faced with the same problem as is the surgeon. I am asked to quote a fee. For some tasks this is satisfactory. In some instances I have been able to quote an hourly fee with a stipulation that not more than a specified maximum will be incurred. In one or two instances there have been problem cases wherein the time used was excessive and the extra time was caused by employees in the organization throwing sand in the gears. Sometimes the only thing I can do is to use the extra time, and figure that I am really working at a lower rate. In one instance an expert in a certain product line was hired by management and asked to improve the product. He made several attempts to improve the product and each one wound up with the product performing essentially as it did initially. He analyzed the pieces and found that each time he had changed one parameter someone had changed another parameter maintaining the performance at the same level. He went back to his mentor and reported. The boss checked this out and found that some engineer had purposely changed the other parameter(s) because he felt he knew more than the consultant. This putting of sand in the gears whether it is deliberate or not can make a short job last forever and can make it impossible to predict what will happen in any specific instance. Is it ethical to have an employee throw sand in the gears and still expect an expert to stick by his or her original estimate of time and cost?

A news item concerns the selling or buying of books that were written by the speaker. The speaker in most of these cases is a legislator. The law says that the speaker cannot accept gross fees in excess of a specific amount in any one year. He or she agrees to speak for no fee, but it is understood that the organization will buy a large quantity of books. Is this an actual sale, a separate business, or a method of getting around a rule which limits speaking fees? Since no legal decisions are being made it must be asked whether this is ethical. If the group who is hearing the speaker and buying the books has an interest in legislation now pending, would you consider this ethical? If only the head of this group has an interest in legislation now pending would you consider this ethical? If the group entering into this agreement were a university with no axe to grind on legislation, would you consider it ethical? If the group entering into such an arrangement was similar to COMMON CAUSE and committed to better government, would you consider it ethical? Under what conditions would you consider it ethical, and under what conditions would you consider it unethical?

On the subject of charging for time, there can be many problems including one in which the expert is one of several working on a problem. One expert renders a report that is circulated to the other experts. There is a fundamental error in this report that a first-year engineering student would recognize. It states that the tension in a wire between two posts is increased by 100 pounds by hanging a 100-pound weight at the center of the span. The expert is called, and he maintains that his report is correct. Is it ethical for the others to go to the lawyer who knows no engineering and point out the error? What if the lawyer says she still believes the report because the writer is an expert?

I was on the witness stand following an engineer from the company we were both representing. The company man had testified that heat rises. The case involved an electric heater which in essence consists of a resistance wire surrounded by a tube and many fins which extend in all directions from the tube. When the heater is connected to an electrical source the resistance wire and fins get hot. The aluminum fins are cooled by the passage of air. This air rises through the spaces between the fins and carries it away from the fins and the resistance wire. The circulating air heats the room. It was considered necessary to disagree with the company engineer and state that heat does not rise. If it did, how is heat gotten from the sun? If it did, why were the fins round and extending in all directions in the form of a circle around the tune and resistance wire core? Heat flowed in all directions from the core and the fins at top and bottom were almost equally effective in providing a method for transferring heat from the core, through the fins, to the air. Fortunately this disagreement did not backfire. The manufacturer won the case. I did better. The expert from the other side referred several cases to me. Was it ethical to disagree with the factory engineer in order to explain what was needed?

The quality control function, like all other functions of an industrial enterprise, is difficult to administer and to operate because it requires careful planning, extreme fairness, and consistency. To operate a quality system properly requires the establishment of precise rules and procedures and the rigorous definition of how they will be applied. In addition it is necessary that the procedures be followed and that deviations from these procedures, which must be implemented, be openly discussed and defended.

An example of the type of problem that might occur is the identification of an undesirable characteristic during the manufacturing process. It is deleterious to the product's use, adversely affects the user's safety, or has an adverse effect on the form, fit, or function of the product. The identification of this deleterious characteristic has occurred when the tools are complete and manufacturing is proceeding. It may be true that the manufacturing group and the inspection and quality people had not been told of this characteristic, but what is to be done now that everyone is aware of it? The rules of the game are changing as the game is being played.

Clearly there is a problem. Production is in process, some product has been shipped, and some may even be in the hands of users. In addition there is a pipeline filled with parts and subassemblies that has this undesired characteristic or effect. The decision as to what is to be done may involve rework, scrap, and even a recall of some units. This is not a decision that should be made by the quality organization alone. It requires a change in rules and decisions, rescheduling, and the expenditure of perhaps large sums. What action would you recommend if this were a critical defect? What action would you recommend if this were a nonconformity that shortened the useful life of the product? For the purpose of your discussion it may be assumed that this is an industry not regulated by the government.

The most important action that the quality organization must take is to assure itself and the company that this is not a sampling peculiarity, but a characteristic not noted or identified earlier, by engineering, manufacturing, sales, or quality.

Sampling variations occur in both attribute and variable sampling and every so often following the statistical nature of data the results are not truly indicative of a change in the process, or of a change of any kind.

In the common attribute sampling procedures a sample is drawn from the lot. The sample consists of a number of units, usually fewer than the number in the lot. The sample is examined for specific characteristics, and if fewer than a predetermined number of units from the sample are nonconforming, the lot is sentenced as acceptable. If the number of nonconforming units found in the sample exceeds the acceptance number, which might vary from as few as zero to a larger number, then the lot is judged as not being acceptable and a procedure agreed upon is initiated.

One of the procedures that may be adopted with a nonacceptable lot is to do a 100 percent inspection and use the remainder after removing the nonacceptable units. Two things may occur during this 100 percent inspection. One is that the inspection finds as many or more nonconforming units than was expected. The other is that the inspection finds no nonconforming units or fewer nonconforming units than expected. When none are found there may be complaints that the sampling inspection was done improperly, and that the selection process did not obtain a truly random sample. The converse could also be said when the inspection turns up many more than expected; but in that case it is understood that the inspection was really needed and no complaint is made. The ethics of when a complaint should be made is not often a consideration.

Similar situations appear on the manufacturing floor when inspectors sort out good and bad parts and supervisors review the bad units and place a number of them into the good group because in their opinion they are usable. This sort of activity results in inspectors flinching in the direction of placing marginal units into the good group. The reverse type of flinching is seldom encountered in factories but it would seem just as likely that the inspectors are erring in one way as in the other.

Sampling errors are most serious when the tests are destructive or when the consequences of failing to identify defects and/or nonconformities are serious.

An interesting case of sampling variability, which received much attention, occurred during one of the draft lotteries in the 1960s. It was not a popular draft. Young men did not want to go to Viet Nam and many families did not want their sons drafted. The draft covered those who reached their 18th birthday during the year. Each day of the year was given a number from 1 to 365, and if your birthday came on a day assigned a low number the likelihood of being drafted was high. If your birthday fell on a day that had been assigned a high number you were relatively certain of not being drafted. The days of the year were assigned numbers by a lottery. Each day of the year was written on a slip of paper and put in a capsule. The capsules were mixed and withdrawn by random selection. The first date drawn, for example, the 15th of March was assigned the number 1 and so on until the last capsule was drawn. The date in the last capsule drawn was assigned the number 365. For leap years there was an extra capsule with the date February 29.

The draft boards called the men in the order of their lottery numbers. If your birth date was assigned the number 1 you were the first to be called. The order of the capsule and date assignment was supposed to be random, but it was noted, shortly after the lottery had taken place, that some months seemed to have disproportionate numbers of days with low numbers and other months had a disproportionate number of days with high numbers. The statisticians proved that this distribution was unlikely. It had a low probability of occurrence. It was surmised that there had been an insufficient mixing of the capsules.

A check into the procedure revealed that this was the case. When January's capsules were mixed a few came apart. When February's capsules were then mixed, even more came apart. The mixers therefore became more gentle, and the mixing might not have been and probably was not as thorough as it should have been. The lottery was much criticized and those that had low draft numbers felt that they had not been fairly treated. Nonetheless the lottery was not changed and continued to be used.

Harold Dodge, of Dodge–Roming Sampling Tables fame and an expert on sampling, commented that if the lottery had resulted in January 1 as the first date drawn, January 2 as the second date drawn and continued in order through the year with December 31st being the last date drawn, no one would have believed that this was a random drawing or that the lottery had been honestly carried out. Almost everyone would have cried foul, especially those who were born during the beginning of the year. Dodge further stated that this peculiar drawing had the same chance of occurrence as any other arrangement, if the drawing was done on a random basis. The results of this lottery were just as likely to occur as any other.

Despite the fact that the drawing was done with somewhat less mixing than people would have liked, were those who conducted the lottery ethical in the discharge of their duty to obtain a random distribution of the dates? Should they have made more effort to prevent clustering? Would a different order of draft been more fair?

The same problems are present in every sampling situation. There will be incidents where 2,000 pieces are presented for acceptance. In some of these situations there will be two nonconforming items. This has been inadvertent and not purposeful. These two units can appear in a sample of 2, 5, 10, 100, 200, and cause the lot to be judged as nonacceptable. They may be marginally out of tolerance for one dimension. The chance that the two units will cause the nonacceptable decision is known as the producer's risk. The chance will be different depending on the sample size and the acceptance number. Likewise with any sampling plan there will be consumer's risk. The producer's risk is the probability that the lot, though really meeting the acceptance criteria as a whole, will be judged nonacceptable in a sampling plan. The consumer's risk is usually defined as the probability that the consumer will find a lot acceptable although it does not meet a criteria of the sampling plan. Often the risks are of the form: a 5 percent risk for the producer of having a lot that meets the contractual agreement of containing less than p percent nonconforming found not acceptable. In a similar vein the consumer's risk might be quoted as a 10 percent risk of accepting a lot as poor as 4 x p percent nonconforming. The producer can improve the probability of the lot being accepted by improving the quality and having a smaller fraction nonconforming.

There are ways to improve the odds. The chances of acceptance vary to a large degree depending on the lot size when the MIL-STD-105 tables are used on an AQL basis. In general, larger sample sizes with larger acceptance numbers provide the opportunity of having better discrimination. Is it ethical to change the lot size to improve the chances that the lot will be accepted when the production quality is markedly better than the AQL? Is it ethical to alter the lot size to improve the chances of the lot being accepted when the production quality is poorer than the AQL?

The subject of ethics is an everyday item in the news especially where government and business are concerned. There are frequent complaints that government representatives are really the delegates of special interests and that these interests obtain favored treatment at the expense of the interests of the general public. There are also complaints in the press that the special interest of certain businesses, such as those on Wall Street, are operating to line the pockets of the executives and operators of the enterprises. In some cases there have been proven allegations resulting in convictions of some members of the business community caught in the act of helping themselves to millions of dollars at the expense of those members of the public who had trustingly believed themselves to be the clients of these executives. Perhaps anyone who is engaged in an operation where he is able to pocket millions, should be defined as an executive, even though he is really a major operator in the business community.

Of late there have been several very large leveraged buyouts. Are the directors of the corporations serving their stockholders with fidelity when they participate or acquiesce to these arrangements, or are they serving their own interests by agreeing to accept "golden parachutes?"

In government at the national, state, and local levels there are all sorts of allegations as to impropriety and lack of ethical conduct. Although these do not come directly under the heading of ethics and quality control practitioners, they come close enough to some situations where the industrial and manufacturing communities could become involved.

The *New York Times* of Sunday August 28, 1988 reported on the following case [34]. Is it ethical for a former county attorney to work as a consultant at a higher salary? The question was raised by legislators who had just passed a law requiring a waiting period of a year between exit from county service to the entrance into consulting contract agreements. The law was to become effective in September, and the attorney was to leave public service on the 30th of September. The attorney resigned in August and was immediately given a contract at a rate exceeding his public service pay. The legislators claimed that the spirit of the law had been contravened.

The hiring of retirees as consultants is not an unusual practice in industry. The hiring of professionals as consultants who leave employers to go into private practice is also countenanced. Their pensions and fees may exceed their former salaries but it is rare that their fees alone do.

Is hiring a public official as a consultant for a government agency at a salary exceeding the public one ethical? Is this a form of favoritism? Is there a justification that one could rationalize for full- or part-time service at the suggested rate? Is the hiring of an individual as a consultant legitimate in these circumstances? Can you rationalize situations where you would and would not recommend such actions?

Another portion of the local ethics law was discussed in the same *New York Times* article. Apparently nominees for government positions would be allowed seven days after nomination to disclose their financial situations and the items they owned that might affect their voting record. Was the seven-day period sufficient for individuals to disclose their financial record? The answers were both yes and no. In most cases individuals would have advance knowledge of their nomination. Some legislators felt that financial disclosures might discourage some people from running for office. Others pointed out that the law would not discourage the dishonest and the crooked. In your opinion would ethics laws discourage meritorious people from running for office? Should candidates for public office be required to file a list of assets? Does owning a share of stock in a company interested in doing business with a government unit disqualify individuals from making decisions for taxpayers? Does owning shares in corporations that are bidding on projects improperly influence engineers in business or quality managers from rendering honest and unprejudiced opinions for the companies which employ them? Do private businesses have the right to request employees to disclose their net worth and what stocks and interests they have outside of their employment? If they do not, should they have such rights? Does the owning of stock have an influence on your decision?

I was a director of an HMO which was and still is considered to be an insurance activity in New York State. Directors were required by the Insurance Department of the State to disclose any interests they or members of their family had in other insurance organizations and the value of such interest. At that time my wife owned a few shares of another health insurance organization which did not operate in our area. I also had some shares of mutual funds which did or did not own shares in for-profit-health-related organizations. Should I have disclosed all of these possibilities?

There is an instance of a judge refusing to disqualify himself in a case involving a corporation in which his wife held some shares of stock. Ethically what would your reaction be if the value of this stock was 1 percent, 10 percent, or 50 percent of the family assets? If that percentage amounted to $1 million would you have a different reaction?

A much more serious question of ethics is the giving and acceptance of gifts, aimed at influencing or swaying the decisions of business people. It was reported in July, 1988 that an MBA graduate, who had been convicted of an influence involvement, lectured students at the New York University School of Business Administration. He told of going along with the schemes of his family business to pay off people who could influence the purchase of pipe by the Washington Public Power Supply System. The pipe was sold at an excessively high price. He referred to himself as a yuppie on the fast track to riches, not questioning the giving of lavish gifts and payments to suppliers and clients calculated to obtain the sales contracts. He questioned the implied rationale of blowing the whistle on family members and stockholders. His attitude was "don't rock the boat." He spent time in prison and was recommending to the students that they observe ethical procedures. Do you become part of the chain that showers gifts and special privileges on those who inspect your products? Do you take them to lunch? Should acceptance of small samples and lunches be banned? Is it improper to offer these amenities?

There are other ethical considerations. In general the law permits lawyers to share in the judgments they obtain. These fees may be in the order of one-third of the award, but in some jurisdictions, they are scaled down as the awards increase. Other jurisdictions limit the amount. In general, engineers cannot work on contingency fees, because they are supposed to present unprejudiced opinions based on scientific findings. The fees engineers earn are based on agreements between them and the lawyers on the scientific and technical problems relating to accidents in which losses were incurred. The losses may be injuries or deaths, or property damage sustained due to some actions by the opposition, their products or their agents. The losses may have been the result of failure to perform or perhaps some untoward event such as fire. The State of New York has decreed a $300 limit on expert fees. For engineers this is considered to be woefully inadequate and equivalent to asking experts to donate their time. Consequently, lawyers may be prevented from obtaining adequate technical advice.

In contrast, *The Expert and the Law*, a publication of the National Forensic Center in Princeton, New Jersey, cites a case where a physician testifying for a participant asked to be paid 20 percent of the plaintiff's gross recovery [35]. This does not seem to be in agreement with the Code of Ethics of the American Medical Association or with the concept that the payment should be in accord with what is normally charged by a physician. It is also possible that under this type of agreement the physician's payment might be nil. Examine this in the light of a quality control expert testifying for a plaintiff who claims to have suffered loss or injury because of the failure of the product to perform in accordance with its intended manner.

An interesting comment on ethical behavior by a corporate entity is found in a 1988 report to stockholders published by Eli Lilly and Company [36]. Lilly declined to furnish a herbicide-defoliant, tebuthiuron, to the U.S. federal government to eradicate coca plants in Peru. These plants are a source for cocaine. Lilly maintained that the product had not been tested and or registered in Peru for this purpose. Should Lilly have taken this attitude with the U.S. Department of State? Was the action ethical or not?

While politicians and business people are accused of being villains and unethical the IEEE *Institute* of August 1988 finds in a survey that the general public sees the clergy, teachers, and engineers as the good guys. This, despite the fact that several televangelists have had bad years with ethics, morals, and general dishonesty. Engineers are reported to have given themselves an even higher rating, believing that they are 98 percent better than did the general public. Lawyers rated themselves 195 percent higher and media personnel were most generous, rating themselves 838 percent better than did the general public!

A publication that regularly discusses the ethics of political activities is *Common Cause*. Archibald Cox, the special prosecutor fired by President Nixon in the so-called "Saturday Night Massacre" is one of its leaders. *Common Cause* points out that there are some activities to limit the revolving door when Department of

Defense personnel leave the military to join a contractor that they had been supervising [37]. There are also rules against the members of the executive department going to work for the people they had contact with while working for the government. However, there are no rules limiting the actions of members of congress. On one day congressional representatives can be voting on bills affecting a large corporation, and the next day be employed by the same corporation. This is not to infer that in all cases these activities are not in the interest of the general public. It does, however, cause some actions to be suspect.

Another activity that *Common Cause* calls suspect is the money contributed to congressional campaigns by PACs. A political action committee is permitted to contribute $5,000 to a congressional campaign, but it can solicit for many times that much from several givers and thereby generate a much larger sum for the campaign war chest.

In another *Common Cause* article, comments were made concerning the activities of professional groups policing the ethical behavior of their members [38]. *Insight* in its May 30, 1988 issue reported that HALT, a legal reform group located in Washington, D. C., gave low grades to groups fielding gripes about lawyers [39]. HALT sent a questionnaire to the groups in each state that police complaints against lawyers. The states of Oregon and Washington earned an A in some categories and 29 of the 34 state groups who responded received a grade of F. HALT was very critical of a gag rule which threatened complainants with jail for contempt of court if they discussed the complaint they had filed with anyone outside of the review group.

Insight also reports in the same issue on a video presentation that is being made to business school students and a board game that discusses ethics [40]. These are particularly important because many MBAs are being accused of considering anything that makes money for them or their company as a correct behavior, and that there has been little consideration of ethics in business.

You should attempt to delineate situations which could occur in your operation where the inspection operation is, or might be, subjected to pressure to accept nonconforming material, or to render judgments which are not in conformance with the true facts. The control of processes by statistical process control is also subject to failure when improper pressure is applied. Can you find any in your operation?

14

Ethics, Standards, Quality, and General Subjects

Ethics is becoming a subject that is receiving more attention. It appears in the newspapers and on the television and radio. Bill Moyers interviewed Michael Josephson of the Josephson Institute for Ethics on a program shown on the Public Broadcasting System channels in September, 1988 [41]. Josephson had been a practicing lawyer who was assigned the task of teaching ethics to lawyers after the Watergate scandal. He subsequently started his own school devoted to ethics. He began his comments by reviewing a recent issue of the *New York Times*. There were five stories concerning ethics on the first page.

These reports concerned individuals in public life accused of cheating and lying to the government, and people on Wall Street who were indicted for the use of schemes to defraud the public and enrich themselves.

Josephson said that there had been very little teaching of ethics to lawyers and others prior to the Watergate incidents. Today, evidence of sleaze, the opposite of ethics, is everywhere. Success is measured not by how ethical you are, but how high your position is, how many people report to you, and how much money you make. The statement is made in yuppy terms—greed is good. Success is measured by how much you acquire.

Corruption is born of the need to win. Corruption is causing the playing field to be tilted so that it favors some rather than providing an equal chance for all. Josephson told the story of two lawyers hiking through the woods who meet a cougar. The story might be told of any profession. One lawyer drops his backpack and the other asks what he is doing. "I am dropping my pack so I can run faster." His friend comments, "You cannot outrun a cougar."

The other answered "I know, but this way I can outrun you." This is an example of a method of causing the odds to tilt.

In the minds of some, rules are restrictions that must be circumvented for one's own benefit. Ethics is doing unto others as you would have them do unto you, and not as they do unto you. Although this is often stated to come from the bible, it predates the bible and can be found in Greek and Chinese philosophies.

There are at least three solutions to all situations. One is to treat it fairly; the others are to lie, and to shade the truth. There are many heroes who self-righteously lie to government agencies, lie to others without question, and sometimes cannot even tell the truth when it is to their advantage.

The professions have a responsibility to teach their members ethical behavior. It sometimes takes considerable courage to turn down the money and act ethically. Josephson was particularly critical of the U. S. Congress which has exempted itself from the act it passed on ethics in government. Josephson was critical of the lobbyists and the large political campaign funds which are amassed by many incumbents.

Josephson felt that the way most people learn ethical behavior is from emulating the actions of those they admire. This is the most productive way to learn. Is it any wonder that the political area is so lacking in ethics when the rewards for failure to observe ethical behavior can be so large?

When examining the quality control profession, ethical behavior is expected, yet in many instances all that is found are quality managers who are little more than inspectors. They are required to pass judgment on the product quality with severe restrictions as to what they can refuse to accept. The most critical case is the situation where inspectors are faced with the decision of either passing the product or not working in the plant any more. The importance of working takes precedence over any considerations of the value of the part, insofar as the specifications are concerned.

When performing an auditing function, the financial, safety, quality system, or quality of product auditor may run into an ethical problem. There are ethics that auditors are asked to observe by their professional background and the society to which they belong. There may be other ethical questions that arise because auditors discover pervasive dishonesty in operations. They also see company employees not doing their tasks correctly and not reporting deviations from procedures and specifications. An example might be one in which a high-strength alloy is necessary to obtain the strength needed for the product to perform reliably and safely. But a cheaper, lower-strength alloy is used with the connivance of engineering, procurement, and management. The purpose is to save a substantial amount of money. What action should the auditors take? What is the ethical action? It has recently been disclosed that a large percentage of bolts have been found to be fraudulently marked as being high strength when they are not.

To recognize and fail to report this deviation from a proper procedure is unethical. Yet presenting the facts may result in other problems. The management of the organization is fully aware of the situation. They do not want it announced to the public or to their customers. What is the auditor to do? Failure to disclose this variation would be dishonest. The Institute of Internal Auditors, Inc. has a code of ethics which exhorts the members to exercise honesty, objectivity, and diligence. Members shall exhibit loyalty, shall avoid entering into any activity which may be in conflict with the interests of their employer, and shall reveal material facts, which, if not revealed could either distort the report of the results of the operations under review, or conceal unlawful practice.

The internal auditor in all cases is faced with the dilemma of how to report and disclose these facts without going outside of the organization and revealing these practices to the public.

Should auditors go outside of organizations? Should they discuss these departures from honest behavior with the organizational management before going elsewhere? Should they report any departures from what they consider to be honest and ethical performance to the outside organization for which they are working, before discussing with the organization they are auditing, to assure themselves that they are reporting the true facts? The Code of Ethics of the Institute enjoins auditors to use all reasonable care to obtain sufficient factual information and also to reveal all material information that is known to them, which if not revealed, could either distort the report or conceal unlawful practice.

The Code of Ethics of the Institute of Internal Auditors also lists the requirements that members shall continually strive for improvement in the effectiveness and proficiency of their services and that they shall not accept a fee or gratuity from a client or customer without the full knowledge of their senior management.

Let it be assumed that the management in the previous case was not aware of the deviation. They are so pleased with the disclosure, which permits them to clean up a situation which could have damaged the company's reputation, that they wish to reward the auditor. Under these conditions one might interpret the acceptance of a new automobile by the auditor and his or her senior manager as a proper gift from a client, and an ethical practice. Would you consider it ethical for your company to offer such gifts, or your auditors to accept such gifts? Why do you have these beliefs?

What if the auditor does not discover the substitution? The management of the audited company is so pleased that they are going to continue with the substitution and the exorbitant profits. They also make an offer of a substantial gift to the auditor and his or her manager. Under what conditions can the gifts be accepted? Is there an ethical solution for the auditors?

It might also be assumed that the audited company is so pleased, in either of the two previously discussed situations, that they offer the auditor employment in the firm at a substantial increase in salary. What are the ethics of acceptance of any of the offers? Why is it proper in one case and not in another?

Another facet of auditing is involved in the Registration of Quality Systems. The practice of registration has grown since the adoption of the International Standards Organization's (ISO) 9000 series of standards: ISO 9000, 9001, 9002, 9003, and 9004. (ANSI/ASQC Standards Q90 through Q94 are the equivalent standards in American English.) These standards define the arrangement whereby the adoption of specific procedures in quality systems may be used as the minimum requirement to comply with standards. Meeting these requirements provides a method for an organization to obtain registration of the quality system as complying with the system standards. In some industries registration will be a requirement for a plant to become a vendor to a specific customer. If the product is to be purchased by a customer, he need not survey the operation if it is possible to obtain assurance that the plant meets the requirements of the system. Registration, following a review by a competent audit crew, is needed to provide this assurance, in most cases. However, once the audit has assured the service that the plant is in compliance with the requirements of the standard, the plant can be registered for the product. Audits by many other organizations may then be avoided. In addition, and by no means a trivial consideration, the procedure allows for product acceptance from a plant with a properly registered quality system, without inspection at the receiving site.

The assumption is made that the plant will be honestly operated and that the management will continue to observe the system requirements. There is a variety of rules and restrictions on plant expansion, movement of the plant, and the question of the compliance of the product with the appropriate standard or specification.

In the ISO system it is proposed that the granting of a registration would be followed up by subsequent, unannounced surveys. Is there a difference between what one would expect to find in an announced and an unannounced survey? Are they both ethical?

One of the most likely fields to find examples of unethical behavior is in the public service sector. Why this is so is not immediately obvious. The truth is that worker no-shows, and all sorts of unethical practices seem to occur in these occupations. There are all sorts of arrangements that come to light in which public funds are appropriated for private good. Some of these are lawful, some unlawful, and some unethical. Interestingly enough there are many instances where the punishment for these transgressions is almost nonexistent, but perpetrators run afoul of the law when they lie under oath and are then convicted of perjury. The general public may not even feel strongly about these transgressions. There have been many instances where a public official has been re-elected to office while serving time for betraying the public trust.

In the management of a quality system, there are long- and short-term objectives. A long-term objective will often be the development of a working total quality control system to the point where there is a complete understanding of the requirements for quality throughout the organization, and the acceptance of the concept that the product designed, manufactured, and sold must not only meet the specifications, but also the expectations of customers. When these conditions have been achieved the work of the quality organization is easier.

Before the total quality system is firmly established, there will be sections and people in the organization who still believe that if the product is able to get by, then that is all that is needed. It is a little like the fastening of bolts on an assembly. Is it acceptable to put the bolts in and turn them down gently so that they look okay? Or is it necessary to torque the bolts in place so that they firmly hold and do not loosen in shipment or in use?

The achievement of acceptable product, in the sense that the product not only performs when it is delivered, but also remains operating in a manner that meets customer expectations, is not a simple task. It also takes some time. The announcement that the product will henceforth be in accordance with the specification and that even minor deviations will not be acceptable is, in some instances, equivalent to shutting down the plant.

Any statement that the manufacturing department will not be permitted to have any lots accepted if, in the sampling of these production lots, any nonconforming units are found, may be equivalent to closing down manufacturing. The cries of anguish from manufacturing may only be exceeded by the cries from sales who have promised customers immediate delivery. Whether such an ultimatum can be sustained is often problematical. The boss may feel that bottlenecking shipments makes it impossible to maintain the plant and the business. Is there an alternative?

If customer dissatisfaction has resulted from the frequency of nonconformances which do not result in total failure to operate, and which are not likely to be hazardous to the users and owners, there are alternatives to the shutdown.

In one plant the units were sold under a long warranty. If the unit proved to be unsatisfactory purchasers could bring them back, or a salesman would visit the site and exchange the units for rebuilt ones that were operating satisfactorily. The complaints came from the sales staff and the customers. There were plainly too many exchanges. A careful analysis of the returned units indicated that the majority of the returns were caused by improper assembly, including improper tightening and positioning of parts, poor soldering, and failure to insert all components. What were the final inspectors doing? The plant was about to start the production of a new model. Charts were set up showing the frequency of various attribute failures. These were posted. This was a union plant, but the union was just as anxious to improve the product quality as was management. At the end of the first week the charts were set up for general display and one employee began to laugh. "How ridiculous, Can you imagine that the screws that hold the case on are not tightened?" Then a moment later he said, "My gosh, that is my job!" A loose screw was never seen again.

There are many other sloppy workmanship habits that were disclosed, but it was found that workmanship was not the primary cause of the product behaving in an intermittent manner. There were manufacturing and engineering improvements to be made. It had been agreed, based on the returns from the sales agents, that the quality level leaving the plant had been at a specific level. It was also agreed that the daily production lots would be sampled after the 100 percent inspection and that they would not be shipped unless the sampling indicated that the lot had fewer than x percent nonconforming, approximately one-fifth of the former level. For almost a week the lots failed acceptance, not only after the first 100 percent inspection and sampling, but after the second and third 100 percent inspection and samplings. The second and third samplings used the same plan and criteria as the first. Finally a lot passed acceptance, and the sales department was notified that it would be shipped. Workers were amazed and pleased to have the sales manager comment that he hoped the initial standards that had been promulgated were not relaxed.

In this instance the delay was permitted, and the system worked because a set of rules had been established before the assembly had been started. It was not a change of rules as the game was being played. As time progressed the product improved, and the many minor problems that contributed to the difficulties were eliminated. The sampling was changed to assure that even better quality went to the field. The ethical practice of full disclosure between departments and the cooperative effort of design, manufacturing, and inspection were required to achieve these breakthroughs to higher levels of quality. The production department workers became enthused about improving the integrity of the product. By careful redesign and procurement and improved manufacturing care, it was found possible to achieve a quality level that dramatically reduced the number, and fraction, of instruments that were unacceptable the first time they were tested. The amount of rework was markedly reduced.

Was it proper to use the same acceptance criteria when sampling a resubmitted lot? This has been previously discussed. However, in this case the cause of failure to pass the acceptance criteria was not the finding of a number of nonconforming units equal to or just greater than the rejection number. The number of failures was much greater.

At a subsequent time the quality people identified a problem in another product which was contributing to field failure. A relatively simple fix was suggested that would, with a high probability, reduce the field failure rate and increase the product's longevity and reliability. A conference was held with the production manager. The solution was simple and not of any consequence insofar as cost was concerned. The production manager suggested that the quality and engineering departments permit him to complete the lot being prepared for shipment today without change and starting on the next day all products shipped would be in accordance with the suggested correction. Today's shipment would represent approximately 1 percent of the total in the field. He reasoned that the 1 percent was trivial and that the correction did not represent the correction of a hazard. Was it ethical to request that shipments not be disrupted? Would it be ethical to allow the shipment of today's product and agree to initiate the change as of the next day? It must be remembered that these units would all be under warranty for a long time and that the company would, in the long run, have to update these units when and if they returned to the plant for servicing.

Some time back the manufacturing line for incandescent lamps in one of the larger lamp works was under study by the engineer in charge. There were units reaching the end of the line that were rejected because the solder joint at the lamp base was not properly made. It was the practice of the factory to collect these and resolder the joint and then ship acceptable bulbs. There was nothing improper about this. The reworked bulbs were equal to the others in all respects. The quality manager decided that there was not going to be any more rework and any bulb reaching the inspection station that was not properly soldered would go into the scrap basket. Was it ethical to scrap these bulbs? What would you suspect occurred as the result of this inspection decision?

The present-day thinking about tolerances is that material close to the tolerance limit is of less value than product that is at center, or target value. This has an effect on the way tolerances should be specified, because a one-sided tolerance would practically never have material on target. This does have relationship to the problem of flinching on tolerance. As has been noted previously, there are at least three kinds of flinching. In one, inspectors classify product that is outside of the tolerance but close, as conforming, because they know that supervisors or reviewers will go through the discarded units and put back into the system those that are marginally bad. Another kind of flinching is the testing and retesting of units that are marginal to see if they will conform. The third is the retesting, rechecking, and discarding of units that are in tolerance but close to the tolerance limit. In each of these cases, inspectors are either trying to avoid criticism or acting, in their belief, in the interest of the company. This is almost a usurping of the authority of the organization's material review board.

The Material Review Board is created in most organizations to examine the possibility of using material that is not in conformance with the product specification. Is it ethical to use material that is not in conformance? If so, under what conditions would the use of such material be proper? Accepting marginal material during a 100 percent inspection may have an adverse affect on the subsequent acceptance sampling procedure. (See previous reference to Grubbs and Coon in Chapter 12.)

In one plant a high-strength bolt was used to fasten bus bars to terminals. The bolt was tightly fastened to a predetermined torque. The procurement and engineering department found a source which was considerably cheaper for this expensive item. The quality manager voiced his concern that the item which had been trouble free would now be replaced by an item on which there was no history. He said he could see no objection to the change of source, but still felt uncomfortable about the resulting contact resistance at the joint and the reliability of the product. It was agreed that a pilot order would be placed and the results carefully watched prior to the 100 percent substitution. The first shipment arrived and was placed in production. The production manager called the quality manager to advise him that the substitution would not be made, because the first effort to torque these bolts to the standard value resulted in the heads of the bolts separating from the shaft. This seems ethical, but there may be other questions that the entire procedure was not accompanied by the necessary safeguards. Was it?

In another instance the laboratory developed a new component for a battery. This component seemed to have several advantages. It was less expensive, took less space, allowed the use of more active material, and had less resistance. Since the battery built with the new material had more capacity and less resistance, it performed better and delivered more total energy. There was some doubt as to what the long-term results would be. There might be some failures of the new item which would result in the total failure of the battery. An agreement was made to manufacture a limited number of units and allow the use of the battery for noncritical applications. Batteries would than be used in the field and experience would be gained. Was this an ethical decision? Things went well for almost year and the pressure began to mount for universal release of this material. Then complaints began to arrive. These batteries were failing for a reason totally unrelated to engineer's suspicions, but the failures were related to a characteristic of the component that had not been foreseen. The reasons for caution had been incorrect, but the use of caution had been correct. On what ethical grounds is it correct to use a limited number of units rather than change the entire production?

In another situation a product was not providing the expected performance and service. A design and production change was found that made substantial improvement. The boss did not want any change until the old material was used up. It is not unusual to establish a changeover that specifies that the substitution will not be made until the material in stock is used. Under what conditions is this ethical?

The establishment of a new standard of performance may in the future require that a changed part be made to improve the performance, endurance,life, or safety of a product. In the automobile and other industries this may result in a recall which may cover a large or limited number of units. It is an expensive and not always efficient procedure as the individuals and companies that receive the recall notice do not all respond. Some of the units that are to be recalled will not necessarily give trouble or be improved. There is a philosophy that, if it isn't broken, don't fix it. Based on this, you must consider the possibility that the inspection and field services may be detrimental.

The number of units included in a recall and the total expense may be markedly reduced if there is some form of serial numbering or date code on each unit indicating the order of manufacture and/or the date the product was made. This coding procedure can often be performed at very low cost. Is it ethical not to code product? What is it that leads you to this conclusion?

The manufacturer of a riding lawn mower developed a safety feature that disengaged the rotating blade from the engine and stopped it rapidly when the driver left the seat or released the accelerator. This safety feature was incorporated in a kit which the owner or the dealer could attach to older units which did not have this feature. This manufacturer did not keep records of owners. He notified the distributors who were to notify the dealers that this kit was available. He did not help advertise the kit which was modestly priced. The installation charge was left to the dealers. Was the manufacturer acting in an ethical manner?

In October 1988 the Sundstrand Corporation agreed to pay the U. S. government well in excess of $100 million as a penalty for excessive and fraudulent charges to government contracts. The auditor who uncovered the falsification of records and other irregularities was Michael S. McConnell, an employee of the Defense Contracts Audit Agency. He reported that upper management tried to stall his investigations, but documents and the low-level engineers who did cooperate led him to his discoveries.

The reports in the *New York Times* [42] and other newspapers indicated that there had been widespread falsification of records and overbilling. These are criminal activities, as well as ethical failures.

It would not be beneficial from this point of view to try and establish just what procedures were followed and why these were not discovered earlier. But, in the event you had been an internal auditor in the Sundstrand Corporation, what would you have done had you discovered these falsifications of records and became aware that this was a company practice that had the approval of those in higher positions? Among the solutions that an internal auditor might select are:

- Saying nothing;
- Reporting it to the supervisor's boss;
- Reporting it to the supervisor's boss and insisting on an answer;
- Reporting it to an outside auditor, or whistleblowing;
- Leaving the company and saying nothing.
- Leaving the company and reporting the matter.

As you consider these and/or additional alternatives, refer to the codes of ethics of internal auditors, engineers, and any others appropriate for you. Do you become a co-conspirator by following any of these alternatives?

Epilogue

The subject of ethics has been discussed in the light of how it may be interpreted in the industrial arena, and with specific reference in many instances, to how the actions of quality personnel maybe interpreted. This discussion considers the work environment more than that outside of work.

What has been observed is that there is a disparity between what some seem to believe practical and proper and what others practice.

The greatest disparity in ethical beliefs seems to be between those who believe there is a public and general responsibility which each engineer ought to assume and those who believe that the first, last, and only responsibility is to one's superiors and employers. There is also a belief among some that, provided the decision results in a profit, it is ethical and proper.

Among the subjects considered is whether there is an ethical difference, or whether the decisions to be made can be different, yet ethical, if one is an engineer, an auditor, a physician, or a lawyer.

If a building is in a state of decay and engineers are auditing the condition, they have a responsibility to act to prevent an injury or loss of life or property. Considering safety, which should they consider first, the protection of property or the safety of people?

If a corporation or a bank is in a state of insolvency, does the auditor have the privilege to keep this a secret from the public so that the insiders can withdraw their funds and leave others to suffer greater losses? Is the situation different for an internal or an external auditor?

Do you as a quality professional have a responsibility to the public, the customer, the management, and to yourself to see that the product leaving the plant at which you work is safe and in conformance with proper specifications; or is your ethical responsibility satisfied by saying that the product is in conformance with specifications? Does it make any difference if you say it is in conformance with specifications when in truth it is not?

Most importantly ask yourself if you performed your job in an ethical way and will be proud to tell your family and perhaps your grandchildren what you did. Will you want to hide this from the public and those close to you?

If this volume helps one person do his or her job better and in a more ethical manner; if it results in fewer injuries or losses from the unexpected and unwanted performance of some product; or if it results in the improved performance of some product made in the United States, and the improvement of our reputation for quality, it will have served a worthwhile purpose.

References

1. The National Commission on Product Safety, *The Final Report of the National Commission on Product Safety* (Washington, D.C.: Government Printing Office, June, 1970), from the Foreword.
2. Laura Sanders, "How the Government Subsidizes Leveraged Takeovers," *Forbes,* Nov. 28, 1988, p. 192.
3. *American Heritage Dictionary of the American Language* (Boston, New York, etc.: Houghton-Mifflin, 1973).
4. D. A. Firmage, trans., *Modern Engineering Practice* (New York: Garland STPM Press, 1980).
5. *The New Columbia Encyclopedia* (New York: Columbia University Press, 1975).
6. ASQC Code of Ethics, reprinted with the permission of the ASQC.
7. *ABET Fundamental Canons*, reprinted with permission of the Accreditation Board for Engineering and Technology, Inc. (ABET).
8. *IEEE Code of Ethics*, Item i, Article IV, copyright © 1979 by IEEE, reprinted with permission of the Institute of Electrical and Electronics Engineers.
9. *ASQC Fundamental Principles*, reprinted with permission of the ASQC.
10. *ABET Fundamental Canons*, reprinted with permission of the Accreditation Board for Engineering and Technology, Inc. (ABET).

11. *IEEE Code of Ethics*, Article IV, copyright © 1979 by IEEE, reprinted with permission of the Institute of Electrical and Electronics Engineers.
12. *ABET Code of Ethics*, reprinted with permission of the Accreditation Board for Engineering and Technology, Inc. (ABET).
13. *ABET Code of Ethics*, reprinted with permission of the Accreditation Board for Engineering and Technology, Inc. (ABET).
14. *IEEE Code of Ethics*, Article I, claus 5, copyright © 1979 by IEEE, reprinted with permission of the Institute of Electrical and Electronics Engineers.
15. Richard Greene, "Policing the Medical Profession," *Forbes*, October 5, 1987, p. 45.
16. James Traub, "Into the Mouths of Babes," *New York Times Magazine*, July 24, 1988, p. 18.
17. No author, "Follow Up" column, "Beech-Nut Convictions Overturned," *Consumer Reports*, June, 1989, p. 354.
18. William Ira Bennett, "Pluses of Malpractice Suits," *New York Times Magazine*, July 24, 1988, p. 31.
19. Michael deCovrey Hinds, "Some Seatbelts Found Inferior—or Lethal," *New York Times*, Oct. 22, 1988, p. 52.
20. Mark S. Frankel, "Dear Colleague," *Professional Ethics Report*, Winter, 1988, vol. I, no. 1, p. 1. This is the newsletter of the American Association for the Advancement of Science, Committee on Scientific Freedom and Responsibility, Professional Society Ethics Group.
21. Arthur E. Schwartz, "Ethics, Law, and Public Policy," *Professional Ethics Report*, Winter, 1989, vol. II, no. 1, pp. 4-5.
22. Arthur E. Schwartz, "Engineers and the Political Process," *Professional Ethics Report*, Winter, 1989, vol. II, no. 1, p. 2.
23. *American Heritage Dictionary of the English Language*, (Boston, New York, etc.: Houghton-Mifflin, 1973).
24. No author, "Follow Up" column, *New York Times*, October 4, 1987, p. 52.
25. N. R. Kleinfield, "The Whistle Blowers' Morning After," *New York Times*, Nov. 9, 1988, sec. 3, p. 1.
26. Stephen H. Unger, "The BART Case: Ethics and the Employed Engineer," *IEEE CSIT Newsletter*, IEEE Committee on Social Implications of Technology, Issue no. 4, Sept., 1973, pp. 6-8.
27. IEEE Bylaws 112.1 and 112.4, copyright © 1979 by IEEE, reprinted with permission of the Institute of Electrical and Electronics Engineers.
28. Gil Courtemanche, "The Ethics of Whistle Blowing," *Internal Auditor*, Feb., 1988, pp. 36-41.

29. The Treadway Commission, *Report of the National Commission on Fraudulent Financial Reporting* (Washington, D.C.: Government Printing Office, Oct., 1987), p. 114.
30. Selwyn Raab, "Inspectors Seized in Extortion Plan in New York City," *New York Times*, Mar. 25, 1988, p. 1.
31. Selwyn Raab, "New York Revising Inspection System to Prevent Graft," *New York Times*, Mar. 26, 1988, p. 1.
32. F. E. Grubbs and H. J. Coon, "On Setting Test Limits Relative to Specification Limits," *Industrial Quality Control*, Mar., 1954, vol. X, no. 5, pp. 15-20.
33. National Highway Traffic Safety Administration, "Lamps, Reflective Devices, and Associated Equipment: Turn Signal and Hazard Signal Flashers," *Federal Register* (Washington, D.C.: Government Printing Office, vol. 37, no. 213, Nov. 3, 1972, 49 CFR Part 571, docket No. 69-18, notice 12).
34. Gary Kriss, "Contract Dispute Will Test County's New Ethics Code," *New York Times,* Aug. 28, 1988, sec. 12, p. 1.
35. National Forensic Center, "Defense Wages Fight for Fair Expert Witness Fees," *The Expert and the Law* (Princeton, N.J.: National Forensic Center, Mar.-Apr. 1986, p. 8.
36. Eli Lilly and Company, "Lilly Declines to Provide Elanco Product for Coca Eradication Program," *Second Quarter Report*, 1988, p. 10.
37. No author, "Alert" Column, *Common Cause Magazine*, May/June, 1988, vol. 13, no. 3, p. 46.
38. Fred Wertheimer, "About Face," *Common Cause Magazine*, May/June, 1988, vol. 13, no. 38, p. 45.
39. No author, Law-Briefing column, "Low Grades for Offices Fielding Lawyer Gripes," *Insight*, May 30, 1988, p. 58.
40. No author, Business-Briefing column, "Examination of Ethics by VCR, Board Game," *Insight*, May 30, 1988, p. 43.
41. Michael Josephson, as interviewed by Bill Moyers, *Bill Moyers, A World of Ideas*, PBS TV show, aired September, 1988.
42. William Glaberson, "Auditor Asserts Contractor Tried to Impede His Search," *New York Times*, Oct. 20, 1988, sec. D, p. 1.

Index